T0191745

Hardware Security and Trust

Nicolas Sklavos · Ricardo Chaves
Giorgio Di Natale · Francesco Regazzoni
Editors

Hardware Security and Trust

Design and Deployment of Integrated Circuits in a Threatened Environment

Editors
Nicolas Sklavos
Computer Engineering and Informatics Department
University of Patras
Patra
Greece

Ricardo Chaves
INESC-ID, IST
University of Lisbon
Lisbon
Portugal

Giorgio Di Natale
LIRMM—CNRS, UMR 5506-CC 477
University of Montpellier
Montpellier
France

Francesco Regazzoni
ALaRI Institute
University of Lugano
Lugano
Switzerland

ISBN 978-3-319-83037-7 ISBN 978-3-319-44318-8 (eBook)
DOI 10.1007/978-3-319-44318-8

Softcover reprint of the hardcover 1st edition 2017

Printed on acid-free paper

This Springer imprint is published by Springer Nature
The registered company is Springer International Publishing AG
The registered company address is: Gewerbestrasse 11, 6330 Cham, Switzerland

Preface

Hardware security is becoming increasingly more important for many embedded systems applications ranging from small RFID tag to satellites orbiting the earth. Its relevance is expected to increase in the coming decades as secure applications such as public services, communication, control and healthcare keep growing.

Concerning all the possible security threats, the vulnerability of electronic devices that implement cryptography functions (including smart cards) has become the Achille's heel in the last decade. Indeed, even though recent crypto-algorithms have been proven resistant to cryptanalysis, certain fraudulent manipulations on the hardware implementing such algorithms can allow extracting confidential information. The so-called side-channel attacks have been the first type of attacks that target the physical device. They are based on information gathered from the physical implementation of a cryptosystem. For instance, by correlating the power consumed and the data manipulated by the device, it is possible to discover the secret encryption key.

New threats have menaced secure devices and the security of the manufacturing process. The first issue is the trustworthiness of the manufacturing process. From one side, the test procedures, which increase controllability and observability of inner points of the circuit, is antinomic with respect to the security. Another threat is related to the possibility for an untrusted manufacturer to do malicious alterations to the design (for instance to bypass or to disable the security fence of the system). The threat brought by so-called hardware Trojans begins to materialize. A second issue is the hazard of faults that can appear during the circuit's lifetime and that may affect the circuit behavior by way of soft errors or deliberate manipulations, called fault attacks.

In 2012, a new COST Action, called TRUDEVICE ("Trustworthy Manufacturing and Utilization of Secure Devices") started in order to cover the above-mentioned topics. COST is an intergovernmental framework for European Cooperation in Science and Technology, allowing the coordination of nationally funded research on a European level. COST increases the mobility of researchers across Europe and fosters the establishment of scientific excellence. COST does not fund research itself but provides a platform for European scientists to cooperate on a particular project and exchange expertise.

In the context of the TRUDEVICE COST Action, we organized in July 2014 a training school in Lisbon, Portugal. This training school aimed at providing theoretical and practical lectures on topics related to hardware security.

The school started with an introductory session on the fundamental primitives for security, from both hardware and software perspectives. This is followed by an introduction on the implementation of attacks and countermeasures, presenting an overview of physical attacks, both passive and active, and some existing countermeasures. Included in this introduction was the description of the evolution of computer technology and cryptography from the ancient past to current days.

Given this introduction, trustworthy manufacturing of integrated circuits was discussed ranging from the implementation of cryptographic primitives to the manufacturing test of secure devices. The fight against theft, cloning and counterfeiting of integrated circuits was also discussed considering both ASICs and FPGAs. Continuing with the trustworthiness of secure devices, lectures on the various forms of attacks were presented, considering fault attacks and differential power analysis and existing countermeasures.

This training school also included a practical session on performing differential power analysis and on how to test random number generation. As a boost to Ph.D. students an extra session to foster the discussion between students also took place.

The editors would like to thank all the contributing authors for their patience in meeting our deadlines and requirements. Moreover, we would like to express a heartfelt appreciation to all the speakers that made possible the training school. Thanks to their great enthusiasm and work that we could have made the TRUDEVICE training school a grand success.

TRUDEVICE training school speakers: Lejla Batina (Radboud University Nijmegen, The Netherlands), Lilian Bossuet (University of Saint-Etienne, France), Jiri Bucek (Czech Technical University in Prague, Czech Republic), Ricardo Chaves (University of Lisbon, Portugal), Amine Dehbaoui (SERMA Technologies, France), Milos Drutarovsky (Technical University of Kosice, Slovakia), Viktor Fischer (Jean Monnet University Saint-Etienne, France), Julien Francq (AIRBUS Defense and Space, France), Ilya Kizhvatov (RISCURE, The Netherlands), Patrick Haddad (STMicroelectronics and Jean Monnet University Saint-Etienne, France), Vincent van der Leest (Intrinsic-ID, The Netherlands), Victor Lomné (ANSSI, France), Nele Mentens (KU Leuven, Belgium), Giorgio Di Natale (LIRMM, France), Martin Novotny (Czech Technical University in Prague, Czech Republic), Paul-Henri Pugliesi-Conti (NXP Semiconductors, France), Francesco Regazzoni (ALaRI Institute of University of Lugano, Switzerland), Nicolas Sklavos (University of Patras, Greece).

This book follows the same structure of the training school in Lisbon. We start with a brief survey hardware implementations of the Advanced Encryption Standard, which is the cryptographic algorithm that we is used as a reference in the forthcoming chapters. The book is then divided into four main sections. The first section covers the implementation attacks, starting from an introduction on fault attacks and side-channel attacks, followed by a practical description of the differential power analysis. The section is completed by some countermeasures against fault- and power-based attacks.

The second section covers the issues of the manufacturing testing of hardware devices implementing cryptographic algorithms. The first chapter is dedicated to the classical manufacturing testing and how it can be exploited in order to retrieve secret data. The second chapter contains a survey of the academic and industrial countermeasures.

The third section is dedicated to hardware trust. The first chapter analyzes trustworthiness of mobile devices, including both hardware and software components. The second chapter focuses on Hardware Trojan detection, particularly critical given the common outsourcing of ASIC manufacture.

The last section covers many aspects of Physically Unclonable Functions (PUFs). The first chapter introduces the topic and presents a survey of existing solutions. The next two chapters covers PUFs implemented on FPGAs using delay elements and ring oscillators.

Patra, Greece — Nicolas Sklavos
Lisbon, Portugal — Ricardo Chaves
Montpellier, France — Giorgio Di Natale
Lugano, Switzerland — Francesco Regazzoni

Contents

Chapter 1
AES Datapaths on FPGAs: A State of the Art Analysis

João Carlos Resende and Ricardo Chaves

1.1 Introduction

The Advanced Encryption Standard (AES) has been the preferred block cipher algorithm for data security since its 2001 approval by the North American National Institute of Standards and Technology (NIST) [19]. In the field of Field-Programmable Gate Arrays (FPGA) technology, prototyping, easy-deployment, and experimentation has become less time consuming, increasing the amount of available options for a custom-made AES implementation. Options in the chosen datapath width, SBox implementation, round (un)rolling, pipelining, etc., result in different trade-offs in terms of throughput, resource usage, and overall efficiency. The main goal for this chapter is to provide the reader with an overall review of the updated state of the art techniques and architectures for AES implementations on FPGA.

This chapter is organized as follows: Sect. 1.2 provides an introduction to the AES algorithm. Section 1.3 insights the most common solutions for the implementation of each AES operation on FPGAs, while Sect. 1.4 explores some architectural choices when implementing the complete AES cipher. Section 1.5 presents a performance comparison of the most updated state of the art, and Sect. 1.6 concludes with some final remarks.

J.C. Resende · R. Chaves (✉)
Instituto Superior Técnico, Universidade de Lisboa/INESC-ID,
Rua Alves Redol 9, 1000-029 Lisbon, Portugal
e-mail: ricardo.chaves@inesc-id.pt

J.C. Resende
e-mail: joaocresende@tecnico.ulisboa.pt

N. Sklavos et al. (eds.), *Hardware Security and Trust*,
DOI 10.1007/978-3-319-44318-8_1

1.2 The AES Algorithm

In the early 1970s, IBM and the NSA (North American National Security Agency) collaborated on designing the Data Encryption Standard (DES), a symmetric block cipher that would become a Federal Information Processing Standard (FIPS) in 1977. It became one the most predominant digital ciphers at the time and the extended scrutiny it was subjected to influenced modern cryptography.

In the 1990s, the exponential increase of computational power rendered DES unsafe has brute force attacks were able to break the cipher in feasible time. This led the USA's National Institute of Standards and Technology (NIST) to open a competition for a new symmetrical encryption algorithm. Several proposals were submitted and discussed, including the Rijndael algorithm [7]. This algorithm allowed several sizes of data and cipher keys, while maintaining a balanced performance between security, resources and computation efficiency, in both hardware and software. In 2001, a subset of the Rijndael algorithm became the Advanced Encryption Standard (AES) [19].

The AES algorithm is a 128-bit block cipher, accepting key lengths of 128, 192, and 256 bits, processed over N rounds, with N equal to 10, 12, or 14 rounds, respectively, as depicted in Fig. 1.1. Each 128-bit (16 bytes) block of plain text is organized column wise, in a 4×4 byte matrix (named State).

After the initial key addition, in which the plain text is XORed with the first 128 bits of the expanded key, the State goes through the several operations. These round operations are: **SubBytes**, where each byte is replaced by another one, which can be implemented by a Look Up Table (SBox); **ShiftRows**, where the rows of the State are left-round shifted; **MixColumns**, where each column of the State is multiplied by a matrix; and **AddRoundKey**, where the entire State is XORed with the corresponding 128-bit Round Key. The decryption process of the AES cipher is performed identically to the encryption, but with the inverse operations. Note that the last round is slightly different since no (Inv)MixColumns operation is performed.

```
AddRoundKey(State, ekey)
for round= 1, round<N, round++ do
  SubBytes(State)
  ShiftRows(State)
  MixColumns(State)
  AddRoundKey(State, ekey[round])
end for
SubBytes(State)
ShiftRows(State)
AddRoundKey(State, ekey[N])
```

```
AddRoundKey(State, dkey)
for round= 1, round<N, round++ do
  InvSubBytes(State)
  InvShiftRows(State)
  InvMixColumns(State)
  AddRoundKey(State, dkey[round])
end for
InvSubBytes(State)
InvShiftRows(State)
AddRoundKey(State, dkey[N])
```

Fig. 1.1 AES encryption/decryption operations

1.2.1 SubBytes Operation

The SubBytes operation is a nonlinear function which replaces one byte by a different predefined byte, given

$$\begin{gathered} b' = SubBytes(b) \Leftrightarrow \\ \Leftrightarrow b'_i = b_i^{-1} \oplus b^{-1}_{(i+4)mod_8} \oplus b^{-1}_{(i+5)mod_8} \oplus b^{-1}_{(i+6)mod_8} \oplus b^{-1}_{(i+7)mod_8} \oplus c_i \end{gathered} \tag{1.1}$$

$$0 \leq i < 8 \;\; ; \;\; c = \{01100011\}$$
$$\{b \bullet b^{-1}\} mod\{M\} = 1 \;\; ; \;\; M = \{100011011\}$$

where b_i^{-1} is the i-th bit of the multiplicative inverse of the input byte b [19]. For efficiency purposes, the SubBytes function is often replaced by an equivalent 256-byte lookup table, designated as SBox. Alternatives to the implementation of this byte substitution considering composite fields also exist [3, 24, 26].

1.2.2 ShiftRows Operation

The ShiftRows operation, as the name implies, is a permutation of the 2nd, 3rd and 4th rows of the State matrix, 1, 2, and 3 positions to the left, respectively. The inverse operation used in decryption, InvShiftRows, is the direct undoing of the former shifting, with the permutations of the same rows 1, 2, and 3 positions to the right. The 1st row of the State matrix does not suffer any changes in either one of these operations. Both operations are depicted in Fig. 1.2.

$$\begin{bmatrix} 00 & 04 & 08 & 0C \\ 01 & 05 & 09 & 0D \\ 02 & 06 & 0A & 0E \\ 03 & 07 & 0B & 0F \end{bmatrix} => ShiftRows => \begin{bmatrix} 00 & 04 & 08 & 0C \\ 05 & 09 & 0D & 01 \\ 0A & 0E & 02 & 06 \\ 0F & 03 & 07 & 0B \end{bmatrix}$$

$$\begin{bmatrix} 00 & 04 & 08 & 0C \\ 01 & 05 & 09 & 0D \\ 02 & 06 & 0A & 0E \\ 03 & 07 & 0B & 0F \end{bmatrix} => InvShiftRows => \begin{bmatrix} 00 & 04 & 08 & 0C \\ 0D & 01 & 05 & 09 \\ 0A & 0E & 02 & 06 \\ 07 & 0B & 0F & 03 \end{bmatrix}$$

Fig. 1.2 AES ShiftRows and InvShiftRows operations

Table 1.1 Byte-by-byte GF(2^8) multiplication

2^n multiplication	Non 2^n multiplication
$01 \times B = (B \ll 0)_{mod(0x11B)}$	$03 \times B = 02 \times B \oplus 01 \times B$
$02 \times B = (B \ll 1)_{mod(0x11B)}$	$05 \times B = 04 \times B \oplus 01 \times B$
$04 \times B = (B \ll 2)_{mod(0x11B)}$	$07 \times B = 04 \times B \oplus 02 \times B \oplus 01 \times B$
$08 \times B = (B \ll 3)_{mod(0x11B)}$	$0F \times B =$ $08 \times B \oplus 04 \times B \oplus 02 \times B \oplus 01 \times B$
	...

1.2.3 MixColumns Operation

In the (Inv)MixColumns operation, each individual column of the State matrix is replaced by its multiplication, in a GF(2^8),[1] through one of the matrices depicted in Eq. (1.2). In order to easily understand the GF(2^8) multiplication used in the AES (Inv)MixColumns, the approach presented in Table 1.1 can be used, namely: when multiplying a byte with a 2^n coefficient, the byte is simply shifted n bits to the left, as depicted in Table 1.1, e.g., $02 \times B = (B \ll 1)$; multiplying with any other coefficient (not a power of 2) requires a composite XOR of the smaller 2^n coefficients, as also depicted in Table 1.1, e.g., $03 \times B = 02 \times B \oplus 01 \times B$. When an overflow occurs on the 8th bit during shifting, the result must be subtracted (by XORing) with the value "0x11B", i.e., reducing it to the irreducible polynomial associated: $x^8 + x^4 + x^3 + x + 1$ [19].

$$\begin{array}{cc} MixColumns\ matrix & InvMixColumns\ matrix \\ \begin{bmatrix} r_{0i} \\ r_{1i} \\ r_{2i} \\ r_{3i} \end{bmatrix} = \begin{bmatrix} 02 & 03 & 01 & 01 \\ 01 & 02 & 03 & 01 \\ 01 & 01 & 02 & 03 \\ 03 & 01 & 01 & 02 \end{bmatrix} \begin{bmatrix} a_{0i} \\ a_{1i} \\ a_{2i} \\ a_{3i} \end{bmatrix} & \begin{bmatrix} r_{0i} \\ r_{1i} \\ r_{2i} \\ r_{3i} \end{bmatrix} = \begin{bmatrix} 0E & 0B & 0D & 09 \\ 09 & 0E & 0B & 0D \\ 0D & 09 & 0E & 0B \\ 0B & 0D & 09 & 0E \end{bmatrix} \begin{bmatrix} a_{0i} \\ a_{1i} \\ a_{2i} \\ a_{3i} \end{bmatrix} \end{array} \tag{1.2}$$

1.2.4 Key Scheduling

The Key Scheduling, also known as Key Expansion, is an inherent subroutine of the AES algorithm. The Key Scheduling is responsible for converting the 128, 192 or 256 bits long cipher key into all the necessary round keys (11, 13 or 15 round keys).

Similar to the AES ciphering procedures, the Key Scheduling is also an iterative process, as shown in Fig. 1.3. It uses the same SubBytes operation as the ciphering process, alongside the specific **RotByte** and **AddConstant** operations. **RotByte** performs a byte-wise left rotation of a 32-bit word. **AddConstant** is the bitwise XOR

[1] Galois Field, or finite field, of order 2^8 [19].

(a) Key Schedule for Encryption

```
KeySchedule(CipherKey){

for i= 1, i< 4, i++ do
   W[i]_32b = CipherKey[i]_32b
end for
for i= 4, i< (4 * 10 + 4), i++ do
   temp_32b = W[i - 1]_32b
   if mod(i, 4)=0 then
      RotByte(temp)
      SubBytes(temp)
      AddConstant(temp, Const[i/4])
   end if
   W[i]_32b = W[i - 4]_32b ⊕ temp_32b
end for
ekey[ 0: ... :10 ]=W[ {0:3}: ... :{40:43} ]
}
```

(b) Key Schedule for Decryption

```
InvKeySchedule(CipherKey){

ekey=KeySchedule(CipherKey)
dkey[0]_128b=ekey[N]_128b
dkey[N]_128b=ekey[0]_128b
for round= 1, round<N, round++ do
   KeyState_128b=ekey[round]_128b
   InvMixColumns(KeyState)
   dkey[N-round]=KeyState
end for
}
```

Fig. 1.3 AES Encryption/Decryption KeySchedule operation for 128-bit keys.

between a 32-bit word and one equally sized constant vector {'Cte'; 0; 0; 0} [19]. The InvMixColumns operation is also used to calculate the round keys for decryption.

At the end of each round of the AES encryption, a Round Key is required. As long as each key is available in its proper time, the Key Scheduling can either be pre-computed or processed in parallel alongside the data encryption. This is not possible during decryption since the process starts with the last calculated round key (as shown in the right side of Fig. 1.3).

Note that the Key Expansion only needs to be performed once for a given cipher key, since it does not depend on the input data. Given that one cipher key is typically used to cipher a large amount of data, the Key Expansion computation does not need to be recomputed often. Different approaches to implement the Key Scheduling are further discussed in Sect. 1.3.8.

1.3 FPGA Techniques for the AES Operations

Most operations of the AES rounds have a mathematical definition behind them, such as the SubBytes, being a nonlinear function, and the MixColumns, being a matrix multiplication in GF(2^8) [7]. Some implementations even change the original mathematical definition for different purposes: speed, resource usage, side-channel protection, etc.; but, regardless of any change, the AES input-output pair has to be maintained [3, 17, 18, 26]. It is also possible to avoid the use of logic in the implementation of the mathematical definition, and simply replace it by equivalent input-output lookup tables [19]. In hardware, this led to two tendencies in implementing the AES operations: through a logical defined function or by addressable memory-based lookup tables.

Logic-based implementations, more common is ASIC designs, use a set of logic gates, placed and routed, to implement the mathematical function that defines a given operation. Typically, logic-based implementations require less resources, but result in slower designs.

Memory-based implementations store the pre-computed result of an operation into a memory-mapped lookup table. These results are then outputted depending on the input value. This type of implementation requires the existence of memory elements, typically resulting in faster designs. This approach is common in software-based implementations [1] but also on FPGAs that have embedded memory blocks [4, 5, 9, 20, 23].

In this section, an overview of the existing state of the art solutions focused on FPGA is presented. The following describes these solutions regarding the implementation of the ShiftRows, SubBytes, MixColumns operations and their respective inverses, for both logic and memory-based approaches. Given the simplicity of the AddRoundKey operation, and of its implementation, it will only be occasionally mentioned when particularly relevant for the resulting structure.

1.3.1 Datapath Width

One of the first decisions when considering the hardware implementation of an AES design, is the datapath bit-width. This dictates how much of the State data is processed at a time: 8, 32, or the full 128 bits per clock cycle iteration. Implementations with 16 and 64-bit datapath designs can also be considered, but are practically nonexistent.

8-bit datapaths [6, 13, 25] require less resources, but also the highest number of iterations (160 or more cycles), and consequently the lowest throughput. Implementations with 128-bit datapaths [2, 4, 10] can process more data in a single cycle (with one or more cycles/round), thus allowing for higher throughputs. Consequently, given the replication of the computation units operating in parallel, higher resource usage is also imposed.

32-bit datapath structures [5, 20, 23] are often consider as the more balanced compromise between performance and resource usage, originating higher efficiency results (throughput/resources).

1.3.2 (Inv)ShiftRows Implementations: Routing, Multiplexing, and Memory Based

As explained in Sect. 1.2.2, the ShiftRows operation requires the shifting of the second to fourth rows of the State matrix. From an implementation point of view, this simply requires that each of the 16 bytes are properly routed to their respective positions. On FPGAs, signal routing is performed by dedicated routing switches,

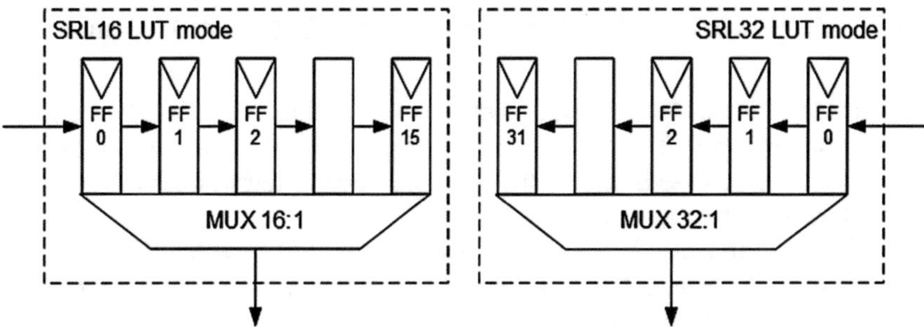

Fig. 1.4 The SRL16 (previous Xilinx FPGAs) and SRL32 (current Xilinx FPGAs) LUT modes

typically not requiring any additional functional logic components. This specific routing is performed when mapping, placing, and routing the structure onto the FPGA. However, ShiftRows and InvShiftRows (used on encryption and decryption, respectively) have opposite shifting directions. Thus the routing path of each operation cannot be shared.

Performing the (Inv)ShiftRows operation through routing is often the preferred choice in several proposed 128-bit datapaths such as Bulens et al. [2] and Liu et al. [17]. However, this implies that a particular implementation can only handle one ciphering mode. With this approach, two AES cores need to be deployed when supporting encryption and decryption, as used in HELION Standard and HELION Fast AES cores [13]. In order to support both encryption and decryption on a single AES design, both routing options need to coexist. If properly designed, and given the similarity of the remaining computations, only minimum multiplexing logic is needed, as presented in Chaves et al. [4].

In smaller datapaths of 32 and 8-bit widths, performing the (Inv)ShiftRows through routing is not viable, since the 16 bytes of the State are not available at the same time. The predominant state of the art solution for the (Inv)ShiftRows in compact FPGA structures is using addressable memory, as introduced in Chodowiec and Gaj [5]. These authors show how a RAM memory can be used to temporarily store the State matrix between rounds, and perform either the ShiftRows or InvShiftRows by properly addressing the writing and reading operations of the consecutive 32-bit columns, or 8-bit cells, of the State [8, 11]. The authors further optimize this byte shift operation by eliminating the need to specify the writing address. This approach is optimized on Xilinx FPGAs using particular LUTs. On these devices, several LUTs have an operational mode called SRL32 (SRL16 in older versions). This mode allows for a single LUT to work as a 32-bit deep shift register with an addressable reading port, resulting in improved resource usage efficiency, as depicted in Fig. 1.4. This approach can be found in 32-bit [5, 20, 23] and 8-bit [6, 25] AES designs.

1.3.3 (Inv)SubBytes Implementations: Logic Versus Memory

Another major implementation differentiation in the state of the art is in the byte substitution operation. These vary from a fine-grained implementation of the byte substitution (Logic-based) [6, 14, 26], to more coarse grained ones using lookup table (Memory-based) approaches [2, 17].

Logic-based structures implement the byte substitution operations by hard-wiring their actual mathematical definition (Sect. 1.2.1) through logic components. If one recalls Eq. (1.1), the SubBytes substitution requires five XOR operations for each bit, but first the multiplicative inverse of the input byte, in the GF(2^8) finite field, needs to be calculated. The problem with the multiplicative inverse is that there is no direct function to calculate it. It is possible to calculate the multiplicative inverse through the Extended Euclidean Algorithm, but this solution is better suited for software rather than hardware [7]. Another approach to compute this multiplicative inverse, more oriented to hardware implementations, is to use Composite Fields [24, 26]. Within logic-based SubBytes implementations, different subsets of Composite Fields can be considered faster, or more compact, or allow for additional security features, than other subsets [3, 18, 22, 26]. The logic-based solution for the InvSubBytes computation is similar to SubBytes, but modifications are still needed.

Overall, logic-based SubBytes implementations are the most area efficient but also the slowest approaches, when compared to memory-based solutions. In a memory-based SubBytes, byte substitution is implemented using a 256-byte lookup SBox table [5, 7, 19]. On FPGAs this can be implemented through the use of multiple FPGA LUTs [2, 17], or even BRAMs [5, 10]. Memory-based approaches can lead to faster circuits at the cost of memory blocks.

On ASIC technology, the decision of using either logic-based or memory-based SubBytes should be carefully analyzed [15]. However, on FPGAs, the use of logic-based implementations has been losing relevancy in comparison to the memory-based counterpart, mainly due to technology improvements. On older or more economical FPGAs, one FPGA LUT can only be configured as a 4-input arbitrary function, with two LUTs per FPGA *Slice*. On more high end FPGAs, such as the Xilinx Virtex 5 and onwards technologies, each *Slice* contains four 6-input LUTs that can be easily combined into a single 8-input lookup table (the exact specification of the AES SBox) with a relatively low latency. If both SubBytes and InvSubBytes operations need to be deployed, either a 9-bit lookup table needs to be considered, or two 8-bit lookup tables multiplexed.

Another easily accessible solution is the use of embedded dual-port memory blocks, BRAMs, that exist within the FPGA. These memory blocks easily allow to store the 2k bits needed for each byte substitution operation.

Implementations that only allow for one ciphering mode often consider the use of LUT-based SBoxes, for shorter clock latency (512 LUTs for 128-bit datapaths [2, 17] and 32 LUTs for 8-bit datapaths [25]). Architectures that allow for both ciphering modes often incorporate pipelined BRAM-based implementations, since they can easily store all tables in their larger memories (8 BRAMs for 128-bit datapaths [10] and two BRAMs for 32-bit datapaths [5]).

1.3.4 Implementing the MixColumns: Logic

After the SubBytes and ShiftRows operations, in the encryption mode, the MixColumns operation is computed by performing a matrix multiplication in GF(2^8). In this operation each 32-bit State column is multiplied by the left matrix of Eq. (1.2), depicting the multiplication coefficients. Similarly to the SubBytes operation, the MixColumns can also be implemented using logic or lookup tables.

In the MixColumns operation each byte is multiplied by a set of four constants ({03}, {02}, {01}, and {01} in the case of encryption). As described in Sect. 1.2.3, the multiplication by 2, in GF(2^8), can be computed by shifting the input value once to the left. If the resulting 9th bit is '1', the entire result has to be bitwise XORed (subtraction in GF(2^8)) by '0x11B', in order to perform the modular reduction. The multiplication by 3 can be achieved by adding the multiplications by 1 (the input value itself) and by 2 (with the addition in GF(2^8) being performed by a bitwise XOR).

To conclude the MixColumns matrix multiplication, the multiplied values are added in GF(2^8) by a XOR tree, as

$$\begin{aligned} r_{0i} &= 02 \times a_{0i} \oplus 03 \times a_{1i} \oplus 01 \times a_{2i} \oplus 01 \times a_{3i} \\ r_{1i} &= 01 \times a_{0i} \oplus 02 \times a_{1i} \oplus 03 \times a_{2i} \oplus 01 \times a_{3i} \\ r_{2i} &= 01 \times a_{0i} \oplus 01 \times a_{1i} \oplus 02 \times a_{2i} \oplus 03 \times a_{3i} \\ r_{3i} &= 03 \times a_{0i} \oplus 01 \times a_{1i} \oplus 01 \times a_{2i} \oplus 02 \times a_{3i} \end{aligned} \tag{1.3}$$

Overall, in a logic-based MixColumns operation, the matrix coefficient multiplications are relatively simple: it requires, for each byte, one 1-bit shift, one 8-bit conditional XOR with the constant '0x1B' to perform the modular reduction (computing ×02), and one 8-bit wide XOR to compute the addition (e.g., $\times 03 = \times 02 \oplus \times 01$). Figure 1.5 illustrates the multiplication of the four coefficients, given one input byte.

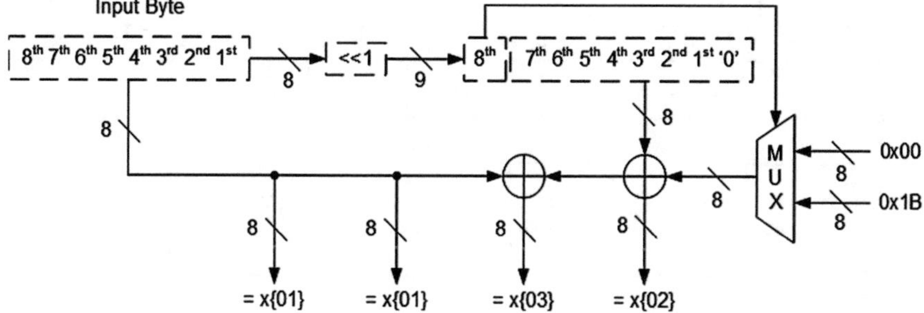

Fig. 1.5 Circuit example for the GF(2^8) encryption multiplication

On a 128-bit datapath, the MixColumns requires a total of 128 7-input functions, or 256 6-input FPGA LUTs. On FPGAs this operation can be performed with relatively low latency, in comparison with the SubBytes stage, as suggested by [2, 5, 10, 17].

On 8-bit datapaths, a single State byte is provided in each clock cycle. As such, the resulting bytes cannot be completed on a single cycle, since each byte resulting from the MixColumns operation depends on four State bytes. Given this, for 8-bit datapaths, registered accumulation can be used. One such approach was first introduced by Hämäläinen et al. [12] for ASIC technology, and later adapted for FPGA by Chu and Benaissa [6]. The resulting structure is depicted in Fig. 1.6.

In this design, the input byte is shifted and XORed in order to obtain the 4 coefficient multiplications ($\{03; 01; 01; 02\}$). The resulting values are then XORed by zero in the first iteration and temporarily stored in four 8-bit registers. In the following cycles, a new input byte suffers the same transformations but is XORed with the previously stored 4-bytes. After 4+1 cycles, one matrix multiplication for one State column is performed. After 16+1 cycles, the entirety of the MixColumns operation can be completed. The issue with this approach [6, 12], is the fact that it requires a 32-bit parallel-to-serial converter, given the 8-bit datapath, as depicted at the bottom of Fig. 1.6.

Instead of performing the 4 coefficient multiplications in parallel, Sasdrich and Güneysu [25] proposed an 8-bit-only accumulative implementation that performs one coefficient multiplication per iteration, as illustrated in Fig. 1.7.

With this approach, a significant area reduction can be achieved by further folding the matrix multiplication and by not needing the parallel-to-serial converter. Additional resources can be saved by preloading a Round Key byte into the register, thus

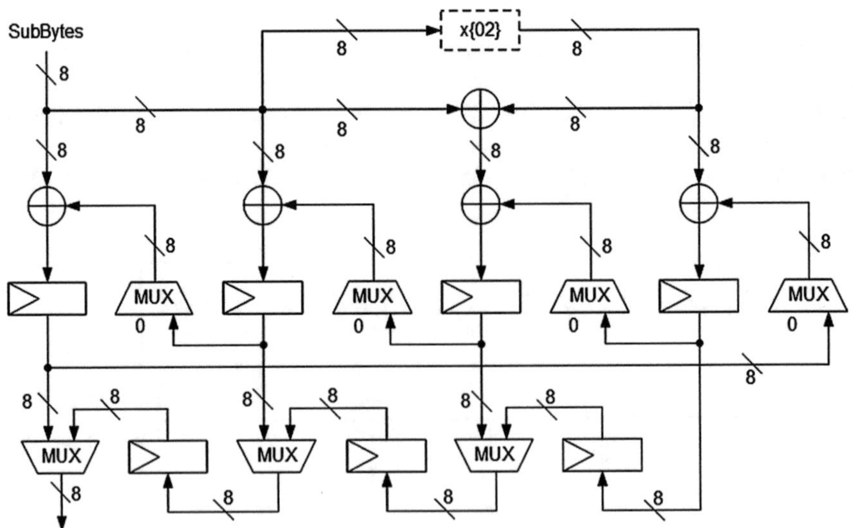

Fig. 1.6 Chu and Benaissa [6] Accumulative MixColumns 8-by-32-by-8 bits

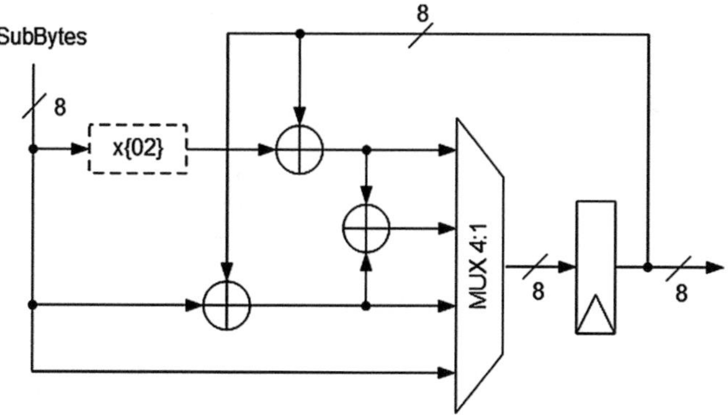

Fig. 1.7 Sasdrich and Güneysu [25] 8-bit Accumulative MixColumns

intrinsically performing the AddRoundKey operation. However, this area compression implies a significant throughput reduction, since it requires 96 clock cycles to complete the MixColumns and AddRoundKey operations. It should be noted that none of these solutions [6, 12, 25] addresses the InvMixColumns operation, which is more complex given the used coefficients ($\{0B; 0D; 09; 0E\}$).

1.3.5 Implementing the InvMixColumns: Logic

The InvMixColumns operation is identical to the MixColumns, but with the coefficients $\{0B; 0D; 09; 0E\}$, resulting in a more complex datapath. The required three modular shifts ($\times 08$; $\times 04$; $\times 02$) and respective XORs (see Table 1.1 and Eq. (1.2)) create a dependency of up to 23 input signals for each bit of the 32-bit matrix multiplication result, as depicted in Fig. 1.8. Because of this complexity, only two state-of-the-art proposals have presented results for architectures with logic-based InvMixColumns [2, 5].

In the single-mode structure presented by Bulens et al. [2], the authors implement the extra required logic for the InvMixColumns (+150 Slices). Chodowiec and Gaj [5] on the other hand, presented a 32-bit datapath that can operate in either encryption or decryption mode. This approach allows to share resources between the two matrices multiplications.

Chodowiec and Gaj [5] realized that, by applying a different, slightly simpler, matrix multiplication over the MixColumns operation, one can compute both the MixColumns and InvMixColumns by sharing resources. Being $c(x)$ and $d(x)$ the polynomials defining the MixColumns and InvMixColumns operations, respectively, and given that

$$c(x)\bullet d(x) = 01 \Leftrightarrow c(x)\bullet d^2(x) = d(x) \qquad (1.4)$$

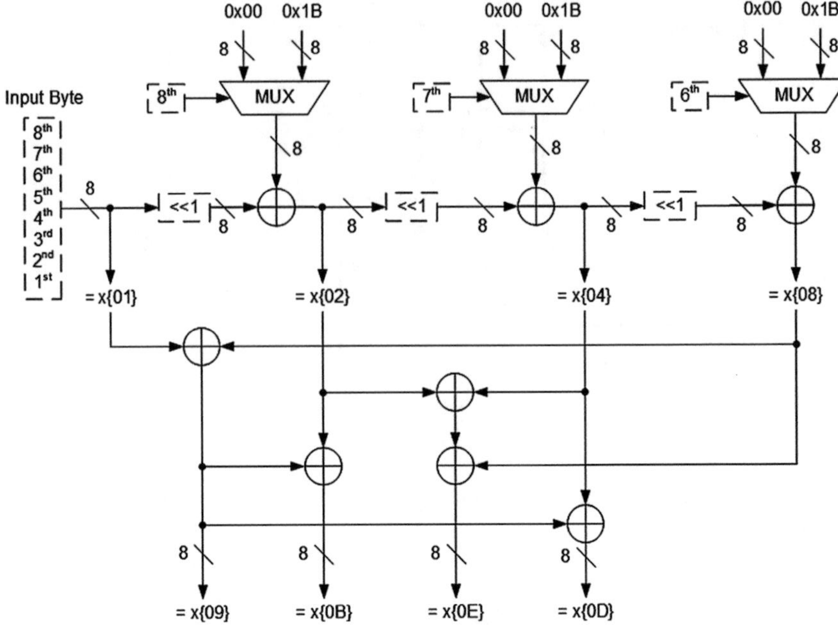

Fig. 1.8 A circuit example for the GF(2^8) decryption multiplication

$$d^2(x) = 04x^2 + 05 \tag{1.5}$$

the InvMixColumns operation can be computed by:

$$\begin{bmatrix} r_{0i} \\ r_{1i} \\ r_{2i} \\ r_{3i} \end{bmatrix} = \begin{bmatrix} 05\ 00\ 04\ 00 \\ 00\ 05\ 00\ 04 \\ 04\ 00\ 05\ 00 \\ 00\ 04\ 00\ 05 \end{bmatrix} \begin{bmatrix} 02\ 03\ 01\ 01 \\ 01\ 02\ 03\ 01 \\ 01\ 01\ 02\ 03 \\ 03\ 01\ 01\ 02 \end{bmatrix} \begin{bmatrix} a_{0i} \\ a_{1i} \\ a_{2i} \\ a_{3i} \end{bmatrix} \tag{1.6}$$

Given this, and by reusing the hardware structure computing the MixColumns, the InvMixColumns operation only requires the additional computational structure computing $d^2(x)$ depicted in Fig. 1.9.

1.3.6 Implementing the (Inv)MixColumns: Memory

Another alternative to implement the multiplication of the coefficients is to map the result into a lookup table. With this option, the multiplication of the input a_{ji} by the coefficients $\{03; 01; 01; 02\}$ or $\{0B; 0D; 09; 0E\}$ are stored into a memory with a 32-bit output. The resulting outputs can then be added (in GF(2^8)) by a tree of XOR

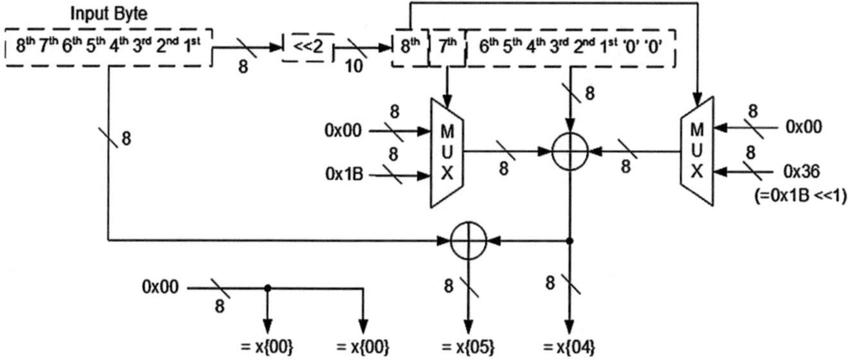

Fig. 1.9 $d^2(x)$ matrix coefficient multiplications of Chodowiec and Gaj [5]

$$TBox(a_{ji}) = \begin{bmatrix} 03 \times SBox(a_{ji}) \\ 01 \times SBox(a_{ji}) \\ 01 \times SBox(a_{ji}) \\ 02 \times SBox(a_{ji}) \end{bmatrix} \quad ; \quad InvTBox(c_{ji}) = \begin{bmatrix} 0E \times InvSBox(c_{ji}) \\ 09 \times InvSBox(c_{ji}) \\ 0D \times InvSBox(c_{ji}) \\ 0B \times InvSBox(c_{ji}) \end{bmatrix}$$

Fig. 1.10 The TBox computation

gates. The mapping of the multiplication coefficients requires $2^8 \times 32 = 8k$ bits of memory for encryption and another 8k bits for decryption.

However, if memory is to be used for the matrix multiplication, the mapped values can belong to $SBox(a_{ji})$ rather than just a_{ji}, i.e., this lookup table can also compute the SubBytes operation. This new table, mapping the byte substitution and the multiplication by the MixColumns coefficients is called TBox, as depicted in Fig. 1.10.

Note that, between the SubBytes and MixColumns operations, the ShiftRows operation should be performed. However, since the byte substitution is the same, independently of the byte position, the ShiftRows can be performed before the TBox computation.

Particularly on FPGAs, the TBox approach is quite recurrent, since it can be easily implemented with embedded memory Blocks (BRAMs on Xilinx FPGAs) acting as 1-byte-by-4-bytes lookup table, as depicted in Fig. 1.10. A single TBox can be stored in any BRAM with at least 8k bits of space and when dual-port access is available in the technology, two substitutions can be performed within the same memory component.

Note that the TBox only provides the GF(2^8) multiplications required for the MixColumns. It does not complete a full matrix multiplication (Eq. 1.2). Only after the input bytes (a_{0i}; a_{1i}; a_{2i}; a_{3i}) have been replaced by TBoxes (resulting in 16 byte parcels) can all the byte outputs be properly "aligned" and XORed to complete the GF(2^8) additions (Eq. 1.3). Proper alignment refers to the different column-wise coefficient sequences found in Eq. (1.3), for each input byte. This can be achieved by having only one type of TBox and afterwards shifting; or by having four different TBoxes, each with a different coefficient rotation ({02; 01; 01; 03}; {03; 02; 01; 01}; {01; 03; 02; 01} and {01; 01; 03; 02}).

When considering the implementation of both TBox and the InvTBox on a single memory block, 9 address bits are used: 8 bits for the input byte and 1 bit for the selection between encryption (TBox) and decryption (InvTBox).

It is also possible to create a LUT-based TBox solution, but doing so is not advised. As Sect. 1.3.3 showed, 32 LUTs per SBox are required, which means 96 to 128 LUTs would be required for each TBox/InvTBox, subsequently surpassing the logic-based solution requirements for large datapaths.

1.3.7 Last AES Round

The last AES round has the particularity of not computing the (Inv)MixColumns operation. Implementations that separate the SubBytes operation from the MixColumns, usually logic-based ones, simply bypass the latter in the last round. However, other implementations, such as the TBox based ones, have the MixColumns operation inherently performed every time. The way the MixColumns operation is bypassed or canceled, in these situations, depends on the details of how the datapath is implemented, as discussed in the following.

The easiest solution is to map into the BRAMs a second set of tables exclusively performing the SBox lookup substitution. This solution is used by Drimer et al. [9] and Resende and Chaves [20], although with slightly different mappings.

In [9], the State bytes are shifted column wise, and each column is fed, one byte at a time, to a specific BRAM port. This means, that in order to properly process all 4 bytes of a column, each 8-bit path needs to be able to access all four rotations of a TBox (plus 4 of the last round TBox), in order to obtain the coefficient alignments of Eq. (1.3). To do this, each of the BRAM's space is occupied with two different rotated tables (+2 last rounds). Logic resources are used afterwards to optionally rotate the 32-bit replaced values, which allows all four rotation types to be obtained, as depicted in Fig. 1.11.

In [20], the memory mappings were improved. The State bytes are shifted line wise, and a full column is fed at each time throughout the four BRAM ports available. This means that every byte that enters in one BRAM port will always need the same TBox (or last round TBox) rotation, halving the memory space required and replacing any extra logic by simple routing, as depicted in Fig. 1.12.

The previous solutions are simple, but still impose additional memory resources, which are not always available. To minimize the memory impact, Rouvroy et al. [23] explores the redundancy of the unitary {01} coefficient in the TBox. A BRAM-based TBox solution is implemented with both TBox and InvTBox tables mapped into memory, as illustrated in Fig. 1.13. For the last *encryption* round, the final substitution is directly obtained from the first unitary MixColumns coefficient $01 \times SBox()$.

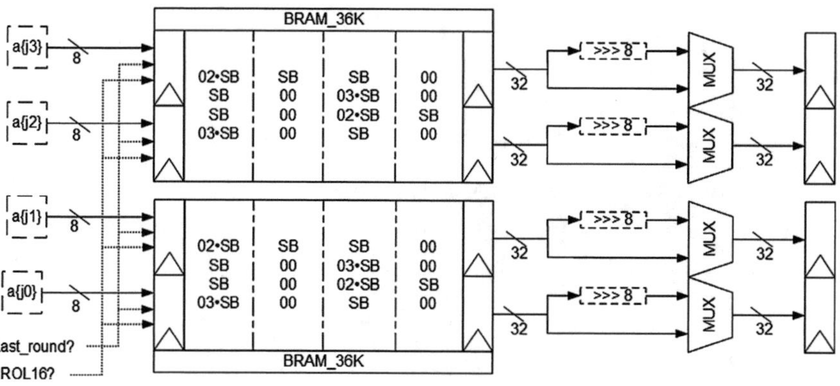

Fig. 1.11 Drimer et al. [9] BRAM-based TBoxes

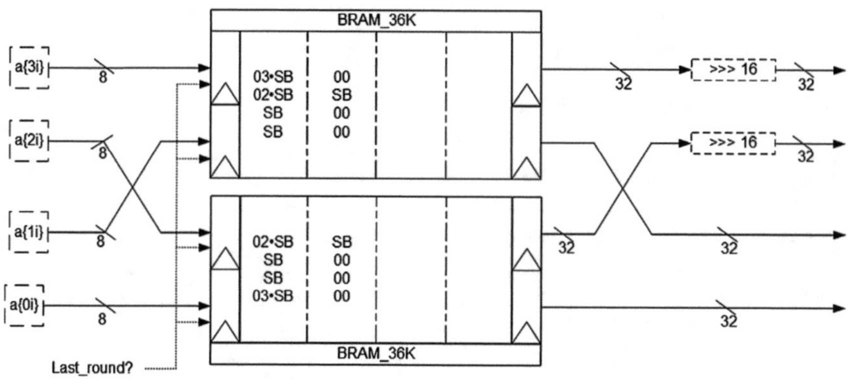

Fig. 1.12 Resende and Chaves [20] BRAM-based TBoxes

The memory space that would be reserved to the second unitary coefficient (01) is modified to contain the value of $InvSBox()$. With this, the InvTBox space is addressed during the *decryption* mid-rounds, while on the last round the TBox space is addressed instead, in order to obtain the $InvSBox()$ value. Any data conflicts are resolved by routing and multiplexing the substitution results. This solution was one of the first in the state of the art considering TBoxes, and is particularly useful when merging the datapath with a Key Scheduling circuit, as proposed in [23]. However, the additional multiplexing logic, placed after the BRAMs, impacts the critical path and consequently the overall performance of the design.

Chaves et al. [4] designed a more elegant solution to cancel the (Inv)MixColumns at the last round. The authors add all the four different matrix coefficients among themselves, using a XOR tree. With this the unitary value can be obtained, as depicted in Eq. (1.7) and Fig. 1.14.

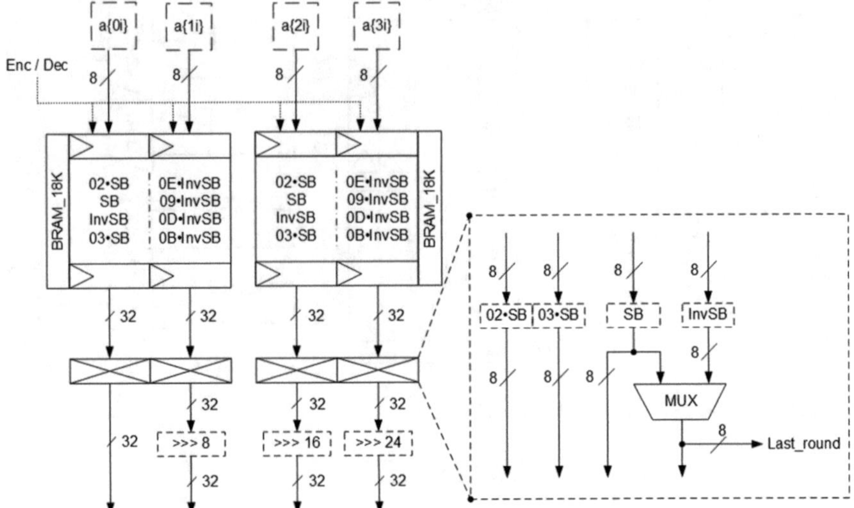

Fig. 1.13 Rouvroy et al. [23] BRAM-based TBoxes

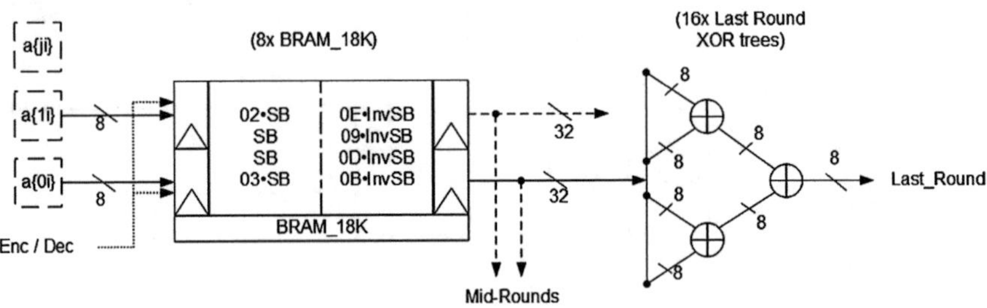

Fig. 1.14 Chaves et al. [4] BRAM-based TBoxes

$$
\begin{aligned}
01 \times b_i &= 03 \times b_i \oplus 01 \times b_i \oplus 01 \times b_i \oplus 02 \times b_i \\
01 \times d_i &= 0B \times d_i \oplus 0D \times d_i \oplus 09 \times d_i \oplus 0E \times d_i
\end{aligned}
\tag{1.7}
$$

This means that a regular TBox substitution can still be performed on the last round, and then canceled by XORing each matrix coefficient multiplication. This solution is easily implemented and its resources are scalable with the datapath width. Moreover, these extra XOR trees can be efficiently separated from the critical path, not impacting the circuit performance [21].

1.3.8 Types of Key Scheduling

When encrypting/decrypting data through AES, several round keys need to be added to the State matrix. If one includes the initial whitening key, a total of 11, 13 or 15, 128-bit round keys are required for the 10, 12, or 14 rounds of AES. As shown in Sect. 1.2.4, the round keys are extracted from the original 128, 192, or 256-bit cipher key defined by the user.

Every time the cipher key is changed, the Key Scheduling must be performed. However, when ciphering multiple data streams, key changes are quite sporadic. This low frequency in which the cipher key and derived round keys need to be updated has led to different approaches to the key scheduling in the state of the art, namely key expansion in parallel with data encryption (On-the-fly) or pre-computed (Off-the-fly); inclusion of dedicated key scheduling logic (On-Chip) or external computation of the key scheduling for higher resource efficiency (Off-Chip). The following briefly analyses these Key Scheduling approaches:

A. On-the-fly versus Off-the-fly: Since data encryption and the expansion of its respective cipher key are both iterative processes, previous works have suggested implementing both to execute in parallel, or at least alternately (On-the-fly). The On-the-fly method has the benefit of requiring very little memory components to store the expanded round keys, since only the most recent one is necessary. On the other hand, since for each encrypted block the round keys always need to be recomputed, either more cycles or more hardware resources are required [6, 10, 17, 25].

Since the cipher key is often maintained throughout the encryption of several data blocks, and if memory components are available, the entire key scheduling can be processed before starting any actual ciphering. In the Off-the-fly approach, all round keys need to be computed and stored, but only once for a given cipher key. At the cost of additional memory, it allows for better throughputs than the On-the-fly solution [2, 5, 23].

As explained in Sect. 1.2.4, the decryption process requires all encrypting round keys to be processed and stored, followed by a post-processing through the InvMix-Columns operation, as depicted in Fig. 1.3. This makes the On-the-fly solution inadequate for decryption, making Off-the-fly Key Schedule the preferred solution when supporting both encryption and decryption.

Good and Benaissa [11] seem to be the only authors to propose a Key Scheduling circuit that can work in both On-the-fly and Off-the-fly modes, operating in the first mode for the first input of the cipher key, and switching to the second mode for as long as the key remains unchanged.

B. On-Chip versus Off-Chip: Given the similarities between the ciphering process and the key scheduling (namely in the SubBytes and the InvMixColumns operations), several implementations have proposed to compactly merge the ciphering process with the key expansion [2, 23]. These solutions, performing the Key Scheduling on the cryptographic engine itself (On-chip), allow to minimize the required resources. However, additional logic is always required [27]. Another option is to perform the Key Scheduling on an external processor (Off-chip) and then load the round keys into a memory component, in the cryptographic engine itself [4, 21, 27].

In the end, the Off-Chip Key Scheduler is preferable when: the cryptographic core is not necessarily autonomous and simply acts as an auxiliary processor and/or; a single cipher key can be kept for large quantities of processable data. The off-chip computation of the key expansion and loading to an auxiliary memory typically yields in more compact and efficient designs [27].

1.4 FPGA Architectures for AES

While the previous section elaborates on the multiple state of the art techniques to improve the implementation of the several AES operations, this section details the architectural options regarding the scheduling of the operations. At this level, the designed decisions are mostly focused on the rolling or unrolling of the loop computation, and in the location and the amount of pipeline stages employed.

1.4.1 Rolled Versus Unrolled Rounds

One of the most direct ways to obtain a trade-off between area and throughput is with round rolling/unrolling.

When unrolling the round computation, multiple rounds of the algorithm are executed in parallel. As such, independent pipeline stages are assigned to each cipher round, as depicted in Fig. 1.15. For this computation to be efficient, data has to be streamed into the pipeline, and the more pipeline stages are placed the faster the overall circuit should run, as described bellow. These approaches are known for imposing higher area demands but, on the other hand, allow for higher throughputs. However, given the data dependency between AES rounds, these approaches can only provide good results if multiple, independent, data blocks are ciphered at the

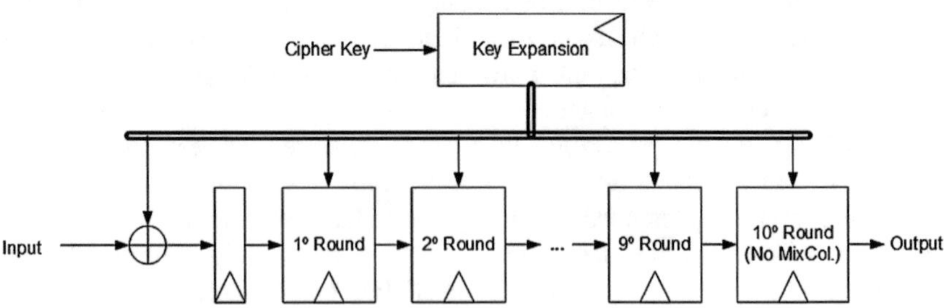

Fig. 1.15 A pipelined unrolled round AES structure

same time. When ciphering in feedback modes (such as CBC) with dependencies between blocks, the throughput improvements cannot be achieved.

Järvinen et al. [16] proposed a fully unrolled pipelined architecture targeting a Xilinx Virtex-II 2000. This solution considers a logic-based implementation requiring four clock cycles to complete each round-stage. Later on, Hodjat and Verbauwhede [14] also designed a four cycles-per-round pipeline structure, logic-based, for the Xilinx Virtex-II Pro. However, these authors also presented a second design that uses a memory-based implementation for the first five rounds (two cycles per stage), and a logic-based implementation for the remaining ones (four cycles per stage).

Regardless of pipeline placement choices, the average throughput across the encryption of a data stream is not directly affected by the increase of pipeline registers in the structure, but by the clock frequency increasing with it. An example of this are two unrolled structures presented by Chaves et al. [4] on the Xilinx Virtex-II Pro. As briefly mentioned in Sect. 1.3.7, both of them use a BRAM-based TBox implementation for all rounds. One structure takes one clock cycle per round, while the second one, with a deeper pipeline, takes three cycles per round. The latter achieves higher clock frequency and throughput values.

The structure presented by Liu et al. [17] updated the AES unrolled structure to the more modern Xilinx Virtex 5, 6, and 7 series. The technological upgrade allowed the authors to use a LUT-based SBOX solution and reduce the pipeline to two cycles per round, while also increasing the clock frequency.

When rolling the architecture, lower hardware requirements are imposed, since only the logic for one round is required. This round structure will process all rounds recursively, taking one or more cycles for each round. Actually, in 32 and 8-bit datapaths, the deployed logic is only able to compute part of the round on each clock cycle. Such datapaths typically allow for relatively small structures, at a cost of lower throughputs [11].

1.4.2 Intra Versus Inter-Pipeline

The clock frequency of a circuit is inversely proportional to the longest propagation delay between two registers of that same circuit. Consequently, the more complex the logic between each pipeline stage, the longer the propagation delay and the lower the clock frequency of the system. As such, the more pipelined the design is, the faster the hardware structure will operate, but more clock cycles will be required to finish a given computation.

In round-based algorithms, such as AES, inter-pipeline refers to the registers that, every clock cycle, store the processed value of one round, and then feed that data to the next round. In rolled round architectures, only one pipeline register is placed between each round logic. The location of these registers can be at the end of the round logic or in between it, such as on the BRAMs computing the TBoxes.

Intra-pipeline refers to the implementation of additional registers between the AES round operations, in order to reduce the critical path and increase clock frequency. Intra-pipeline can exist in either unrolled and rolled round structures.

Strongly unrolled round architectures should always aim to have as much pipeline registers as possible to achieve the highest clock frequency, as their throughput performances are not affected while streaming independent data blocks [14, 17]. For rolled architectures, however, a trade-off between number of cycles and their latency needs to be considered when planning an AES pipelined structure.

In rolled structures, several implementations with 1, 4, or 8 cycles per round have been presented [4, 5, 9, 20, 23], with lower to higher clock frequencies, respectively.

1.5 State of the Art Metrics

The previous sections depict the several design options, proposed in the state of the art, regarding the implementation of the AES on FPGAs.

In this section, a comparative study of the most relevant state of the art structures is presented. Note that not all structures mentioned above are compared, as their original results are somewhat outdated, such as [5, 14, 23] and the unrolled datapaths of [4].

An overview of the architectural features and performances of these structures is depicted in Table 1.2, with their designs grouped by the datapath width.

Regarding the achieved throughput, the presented values define the average rhythm at which each circuit processes the input blocks. In deeper pipelined structures higher throughputs can be achieved, but only if multiple blocks are processed simultaneously. When considering a single data stream in feedback modes, such as CBC, these structures cannot be efficiently used due to the data dependency between blocks. In Table 1.2 the throughput values are depicted as presented by their authors, often considering independent data blocks.

Efficiency wise, we consider the use of the throughput per Slice metric (Throughput/Slice). This metric can be contested as a biased measurement, since it does not take into account other FPGA modules such as BRAMs or DSPs. However, given the difficulty in extracting equivalency values, this is herein used as the efficiency comparison metric.

Regarding the key expansion, Table 1.2 differentiates On-chip (Y) and Off-chip (N) Key Scheduling circuits.

Most of the state of the art proposes architectures capable of performing only AES encryption, and often neglect details regarding the decryption operation. There are two main reasons for this. The first reason is detailed in Sect. 1.2.3 regarding the added complexity of the InvMixColumns operation on logic-based solutions. The second reason regards the structures that include On-the-fly Key Scheduling logic, as introduced in Sect. 1.3.8, since it is extremely inefficient for decryption, given the inverted order in which the round keys are supplied. Note that several ciphering modes only require the existence of encryption, such as Counter and CCM modes.

Table 1.2 Performance comparison

					Resources				
Datapath		Operation[a]	KeySch.	Device	Slices	BRAM	Frequency (MHz)	Throughput (Gbps)	Efficiency (Mbps/Slice)
8-bit	Chu [6]	E	Y	xc3s50-5	184	0	45.6	0.036	0.20
				xc6slx4	80		72.6	0.058	0.72
	Sasdrich [25]	E	Y	xc6slx4	**21**	0	105	0.009	0.09
	Helion [13]	E+D	Y	xc6s.-3	90	0	n.a.	>0.044	>0.49
	Tiny			xc5v.-3	94			>0.077	>0.82
32-bit	Drimer [9][b]	E	N	xc5vsx50t-3	107	2+1	**550**	1.76	16.45
					212				8.30
	Resende [20]	E	N	xc3s4000-5	142	2+1	179.5	0.575	4.05
				xc5vlx30t-3	**70**		530	1.696	**24.22**
				xc6vlx75t-3	**51**		486	1.555	**30.49**
	de la Piedra [8][b]	E	Y	xc7a200t	80	11	91.5	0.195	2.44
	Helion [13]	E	Y	xc5v.-3	155	0	n.a.	>0.84	>5.42
	Standard			xc6v.-3	129			>0.95	>7.36
128-bit	Chaves [4]	E+D	N	xc5vlx30-3	407	8+2	189.5	2.427	5.96
	El Maraghy [10]	E	Y	xc5vlx50	303	8+2	425	1.327	4.38
	Bulens [2]	E	Y	xc5v	400	0	350	**4.167**	10.41
		D			550				7.57
	Liu [17]	E	Y	xc5vlx85	3579	0	360	**46.093**	12.88
				xc6vlx240t	3121		501	**64.128**	20.55
				xc7vx690t	3436		516.8	**66.10**	19.20

[a]Encryption and/or Decryption
[b]Implementations using DSP blocks

The depicted HELION AES cores [13] are commercial intellectual properties, which specific details are not publicly known, and are used here for comparison as market products.

Datapaths with only 8-bit widths are relatively uncommon since their throughputs are normally below 100 Mbps, with wider structures easily surpassing the 1 Gbps mark. This also affects their efficiency. They have, however, a better potential to require less resources.

In 8-bit Datapaths, the work in [6] only requires 80 Slices on a Xilinx Virtex 6, working at 72 MHz, achieving throughputs of 58 Mbps and an efficiency metric of 72 Kbps/Slice. The MixColumns is performed by a 32-bit parallel module and the SubBytes operation is logic based. The design proposed in [25] considerably reduces the amount of resources by using a single 32 LUT-based SBox proposed in [2], while also folding the MixColumns logic for 8 bits only. It is the smallest AES structure presently conceived, with only 21 Slices and a clock frequency of 105 MHz. Throughput and efficiency are inevitably dropped, as expected from an 8-bit datapath, to 9 Mbps and 90 Kbps/Slice.

Although slightly wider than their 8-bit counterparts, the 32-bit datapaths can still offer extremely compact solutions at much higher performances. This is due to the fact that 32-bit widths can take better advantage of several FPGA technology features, such as BRAMs [9, 20] and DSPs [9].

The 32-bit compact structure proposed in [9], allows for a throughput up to 1.76 Gbps at a cost of 107 Slices. This is a TBox-based structure using BRAMs and 4 DSP blocks. The DSP blocks are used to implement the XOR operations, instead of regular Slices. This approach allows for an efficiency of 16.45 Throughput/Slice, achieved for two parallel block streams. Note that DSPs are Xilinx FPGA dedicated arithmetic components and are not accounted for the efficiency metric herein considered. Without the use of DSPs, 212 Slices are needed instead, resulting in an efficiency of 8.30 Mbps/Slice.

The work of de la Piedra et al. [8] extends the use of DSPs. Since DSPs are capable of performing XOR operations with constants, the authors decided to implement them on a logic-based MixColumns. SBox lookup operations and temporary State storage is performed exclusively by BRAMs. This leads to the small amount of needed Slices (80), mostly used in the Key Schedule and Control Unit. However, 11 BRAMs and 16 DSPs are required as a consequence. With this type of resources it is hard to compare the 2.44 Mbps/Slice efficiency value with the remaining state of the art, but remains as a viable alternative, if FPGA Slices are required for other operations.

The work presented in [20] improves upon the two block stream computation of [9]. This design does not use DSPs, allowing for a simple scheduling, and uses the SRL32 LUT-based shifter [5], instead of the large shift register used in [9]. This leads to the best Throughput/Slice efficiency value in the state of the art, 24 Mbps/Slice on a Virtex 5, with a cost of 70 Slices and 3 BRAMs [20], and a throughput of 1.7 Gbps.

Increasing the datapath width allows for better throughputs, but the extra resources impact the area efficiency. The 128-bit rolled datapath structure presented in [4] computes a round on a single clock cycle, achieving a throughput of 2.4 Gbps with an efficiency of 5.96 at a cost of a higher BRAM usage (10 BRAMs).

Bulens et al. [2] designed a 128-bit datapath with LUT-based SBoxes, off-the-fly Key Scheduling and four pipeline stages. This allows to process with a 4.1 Gbps throughput and a 10.4 Mbps/Slice efficiency mark, but only if four different data blocks are being processed in parallel. The datapath presented by El Maraghy et al. [10] has similarities with [2], but the SBoxes are BRAM-based, and the on-the-fly Key Schedule only allows for 1 single 128-bit block to be processed at a time at 1.3 Gbps, with an overall 4.38 Mbps/Slice area efficiency.

Finally, the fully unrolled AES architectures can achieve the highest speed performances for non-feedback streams. The specific solutions for each AES round operation are very similar to the ones presented in the 128-bit rolled round structures: Routed ShiftRows, LUT-based SBox, and logic-based MixColumns, but their inter-pipeline feature suits them for a complete different area of applications. Liu et al. [17] has presented what may be the AES implementation with the highest throughput on the FPGA state of the art, including results for the Virtex 7 FPGAs. With two clock cycles per round, it can reach throughputs from 46 Gbps up to 66 Gbps, and from 12 Mbps/Slice up to 20 Mbps/Slice in efficiency values, depending on the technology. However, the high resources cost, 3121 to 3579 Slices, makes this architecture unsuitable for small embedded devices. Flexibility is also disregarded, as their throughput and efficiency performances can drop by a factor of $\sim$20 if feedback ciphering modes are required.

In the works presented in [9, 20], the authors discussed the difficulty in obtaining high clock frequencies. As explained in Sect. 1.2, the ShiftRows data dependency dictates that 128, 32 and 8-bit datapaths need to complete each AES round in one [4], four [5, 23], and eight cycles, respectively, in order to prevent empty cycles with no computations. However, due to the technology limitations, the minimal amount of cycles per round cannot be reached without affecting the clock frequency. Drimer et al. [9] and Resende and Chaves [20] have stated that the best way to achieve high clock frequency in 32-bit rolled AES datapaths is to create an eight-cycle pipeline for two independent data blocks.

1.6 Conclusion

Since the introduction of the AES, several optimization features have been proposed towards improving the implementation of this algorithm, exploiting the particularities of the FPGA technology. These optimizations consider the several AES operations and the overall computational flow. This chapter provides the reader with an overview of the state of the art techniques used in the implementation of AES on FPGAs.

Datapath widths are herein discussed, ranging from the 8-bit rolled structure proposed by Sasdrich and Güneysu [25] trading speed for compactness, being the smallest architecture with only 21 Slices; the 128-bit unrolled pipeline, proposed by Liu et al. [17], achieving the best streamed throughput of 66 Gbps; and the structure with the highest Throughput/Slice efficiency on a more balanced 32-bit rolled architecture by Resende and Chaves [20], achieving an efficiency metric of 30 Mbps/Slice. Particular FPGA optimizations are also analyzed such as the LUT-based addressable shift register solution for the ShiftRows operation, from its first introduction by Chodowiec and Gaj [5], to its most recent usage in [6, 20, 25]. Logic versus memory-based solutions are also compared, regarding the implementation of both the SubBytes and the MixColumns, with memory lookup tables (such as BRAM-based TBoxes), particularly in structures where both encryption and decryption are necessary.

Additional insights, such as the last round exception or pipeline distribution, are also herein discussed, presenting the most significant contributions in the implementation of the Advanced Encryption Standard on Field-Programmable Gate Arrays.

Acknowledgements This work was partially supported by the ARTEMIS Joint Undertaking under grant agreement no 621429, the TRUDEVICE COST action (ref. IC1204) and by national funds through Fundação para a Ciência e a Tecnologia (FCT) with reference UID/CEC/50021/2013.

References

1. Bos JW, Osvik DA, Stefan D. Fast Implementations of AES on various platforms. IACR Cryptology ePrint Archive, 2009. p. 501.
2. Bulens P, Standaert FX, Quisquater JJ, Pellegrin P, Rouvroy G. Implementation of the AES-128 on virtex-5 FPGAs. In: Progress in cryptology–AFRICACRYPT. Springer; 2008. p. 16–26.
3. Canright D. A very compact S-box for AES. Springer; 2005
4. Chaves R, Kuzmanov G, Vassiliadis S, Sousa L. Reconfigurable memory based AES co-processor. In: 20th international parallel and distributed processing symposium IPDPS, IEEE; 2006. p. 8.
5. Chodowiec P, Gaj K. Very compact FPGA implementation of the AES algorithm. In: Cryptographic hardware and embedded systems-CHES. Springer; 2003. p. 319–33.
6. Chu J, Benaissa M. Low area memory-free FPGA implementation of the AES algorithm. In: 2012 22nd international conference on, field programmable logic and applications (FPL), IEEE; 2012. p. 623–6.
7. Daemen J, Rijmen V. AES proposal: Rijndael, 1999.
8. De La Piedra A, Touhafi A, Braeken A. Compact implementation of CCM and GCM modes of AES using DSP blocks. In: 2013 23rd international conference on, field programmable logic and applications (FPL). IEEE; 2013. p. 1–4.
9. Drimer S, Güneysu T, Paar C. DSPs, BRAMs, and a pinch of logic: extended recipes for AES on FPGAs. ACM Trans Reconfig Technol Syst. 2010;3(1):3.
10. El Maraghy M, Hesham S, Abd El Ghany MA. Real-time efficient FPGA implementation of aes algorithm. In: 2013 IEEE 26th international, SOC conference (SOCC). IEEE; 2013. p. 203–8.
11. Good T, Benaissa M. AES on FPGA from the fastest to the smallest. In: Cryptographic hardware and embedded systems—CHES. Springer; 2005. p. 427–40.
12. Hämäläinen P, Alho T, Hännikäinen M, Hämäläinen TD. Design and implementation of low-area and low-power AES encryption hardware core. In: 9th EUROMICRO conference on, digital system design: architectures, methods and tools, DSD 2006. IEEE; 2006. p. 577–83.
13. Helion. AES CORES. http://www.heliontech.com/aes.htm.

14. Hodjat A, Verbauwhede I. A 21.54 Gbits/s fully pipelined AES processor on FPGA. In: 12th annual IEEE symposium on, field-programmable custom computing machines, 2004. FCCM, IEEE; 2004. p. 308–9.
15. Hodjat A, Verbauwhede I. Area-throughput trade-offs for fully pipelined 30 to 70 Gbits/s AES processors. IEEE Trans Comput. 2006;55(4):366–72.
16. Järvinen KU, Tommiska MT, Skyttä JO. A fully pipelined memoryless 17.8 Gbps AES-128 encryptor. In: Proceedings of the 2003 ACM/SIGDA eleventh international symposium on field programmable gate arrays. ACM; 2003. p. 207–15.
17. Liu Q, Xu Z, Yuan Y. A 66.1 Gbps single-pipeline AES on FPGA. In: 2013 international conference on, field-programmable technology (FPT). IEEE; 2013, p. 378–81.
18. Mentens N, Batina L, Preneel B, Verbauwhede I. A systematic evaluation of compact hardware implementations for the Rijndael S-box. In: Topics in cryptology—CT-RSA 2005. Springer; 2005. p. 323–33.
19. NIST: FIPS 197. Advanced encryption standard (AES). Fed Inf Process Stand Pub. 2001; 197:441–311.
20. Resende JC, Chaves R. Compact dual block AES core on FPGA for CCM protocol. In: 25th international conference, field-programmable logic and applications, FPL 2015 London, September 2–4. IEEE; 2015.
21. Resende JC, Chaves R. Dual CLEFIA/AES cipher core on FPGA. In: Sano K, Soudris D, Hbner M, Diniz PC, editors. Applied reconfigurable computing, lecture notes in computer science, vol. 9040. Springer International Publishing; 2015. p. 229–40.
22. Rijmen V. Efficient implementation of the rijndael S-box. Dept. ESAT. Belgium: Katholieke Universiteit Leuven; 2000.
23. Rouvroy G, Standaert FX, Quisquater JJ, Legat J. Compact and efficient encryption/decryption module for FPGA implementation of the AES Rijndael very well suited for small embedded applications. In: Proceedings of ITCC 2004, international conference on, information technology: coding and computing, vol. 2. IEEE; 2004. p. 583–7.
24. Rudra A, Dubey PK, Jutla CS, Kumar V, Rao JR, Rohatgi P. Efficient rijndael encryption implementation with composite field arithmetic. In: Cryptographic hardware and embedded systems CHES. Springer; 2001. p. 171–84.
25. Sasdrich P, Güneysu T. Pushing the limits: ultra-lightweight AES on reconfigurable hardware. In: Workshop on trustworthy manufacturing and utilization of secure devices. TRUDEVICE; 2015.
26. Satoh A, Morioka S, Takano K, Munetoh S. A compact rijndael hardware architecture with S-box optimization. In: Boyd C, editor. Advances in cryptology ASIACRYPT 2001, vol. 2248., Lecture notes in computer science. Berlin, Heidelberg: Springer; 2001. p. 239–54.
27. Sklavos N, Koufopavlou O. Architectures and VLSI implementations of the AES-proposal Rijndael. IEEE Trans Comput. 2002;51(12):1454–9.

Chapter 2
Fault Attacks, Injection Techniques and Tools for Simulation

Roberta Piscitelli, Shivam Bhasin and Francesco Regazzoni

2.1 Introduction

Embedded systems pervaded our live since few years. The applications where they are used are often safety critical, such as public transports or smart grids control, or handle private and sensitive data, such has medical records or biometrics information for access control. This trend is expected to even increase in the near future, when a large amount of embedded devices will be connected to the so called Internet of Things (IoT). If, on one side, the level of interoperability and connectivity which will be reached by the objects in the IoT will allow to offer a large variety of services, to increase the efficiency and to reduce the costs, on the other side, the envisioned applications require the device to include security functionality to guarantee the confidentiality of the processed data and the security of the overall infrastructure.

Designers anticipated these needs by augmenting several devices with state of the art cryptographic primitives: embedded processors included instructions to quickly encrypt and decrypt data and a number of low-cost accelerators were designed to boost the performance of secure protocols implemented in wireless sensor nodes. However, robust and mathematically secure cryptographic primitives are not sufficient to guarantee the security of embedded devices. In the past, cryptographics algorithm have been conceived to be robust only against mathematical attacks: their structure is realized to resist, among other, to linear and differential cryptanalysis, they were requested to resist brute force attacks, also considering the progress of the

R. Piscitelli (✉)
EGI.eu, Science Park 140, Amsterdam, The Netherlands
e-mail: roberta.piscitelli@egi.eu; roberta.piscitelli83@gmail.com

S. Bhasin
Temasek Labs@NTU, 21 Nanyang Link, Singapore 637371, Singapore
e-mail: sbhasin@ntu.edu.sg

F. Regazzoni
ALaRI-USI, via Buffi, 13, 6900 Lugano, Switzerland
e-mail: regazzoni@alari.ch

N. Sklavos et al. (eds.), *Hardware Security and Trust*,
DOI 10.1007/978-3-319-44318-8_2

technology, and the hardness of the computational problem involved should have been capable of guaranteeing long-term security.

The situation changed in the last to decades, with the advent, the rise and the develop of a novel form of attacks, called *physical attacks*. These attacks, instead of addressing the mathematical structure of the algorithm, try to extract information about the secret key exploiting the weaknesses of the implementation of the algorithm itself. To recover the secret data, the adversary can exploit either an additional information leaked by the device during the computation (for instance the power consumed by the device) or can actively induce an anomalous behavior capable of leaking secret information.

Physical attacks are particularly dangerous for embedded systems, as they are, potentially, "in the hand" of the adversary, which thus has the whole control over them. Physical attacks are usually divided into two classes: *passive attacks* and *active attacks*. Among the first ones, the most notable one is power analysis [1], in which the adversary measures the power consumed by certain number of encryptions computed using a known plaintext, makes an hypothesis on a small portion of the secret key, and used the previously collected power traces to verify the correctness of the hypothesis. Nevertheless, the time [2] needed to complete an encryption, the electromagnetic emission [3] of a device or even the photons emitted by transistors [4] were successfully used to recover secret data.

During active attacks, the adversary does not limit himself to the observation of information leakage but actively tampers with the target device. The most common form of active attack is fault injection. In this attack, the adversary forces the device to perform erroneous operations and he exploits the relation between the correct results and the incorrect ones to infer the secret key (or to significantly reduce the possible key space). Fault injection consists of two parts: the first is the injection of the fault into the device, in which the target device is induced into an anomalous behavior, the second is the attack itself, in which the erroneous output is used to extract secret information.

Fault injection attacks are extremely dangerous because they require a limited amount of time to be carried out and because they were proven to be effective even when performed with an extremely inexpensive equipment. Barenghi et al. [5], for instance, showed how, by underpowering an ARM9 embedded processor, it was possible to induce a number of errors sufficient to successfully attacks software implementations of the AES and the RSA algorithms. A similar approach was used also to attack an ASIC implementation of the AES algorithm.

Robustness against fault attacks is usually evaluated in laboratories, where a manufactured device is tested by mounting a number of known attacks. However, this is not the best solution for designers which needs to timely apply the proper countermeasure against these attacks. Even if the final prove of resistance can be obtained only with the direct evaluation of the manufactured device, it would be more effective to have an initial exploration of the resistance against fault attacks at design time. This, however, requires to have tools capable of simulating the behavior of a device under attack, at the needed resolution, and to have a methodology to compare different countermeasures.

This paper addresses the problem of fault attacks. First, we survey the most common methods used to inject the faults, highlighting the potentialities of the method and its cost. Then we summarize the type of fault attacks previously presented in literature, finally we introduce the design tools which can be used for simulating fault attacks and we discuss to which extend they are suitable for evaluating the sensitivity of a device against fault attacks.

2.2 Fault Injection Techniques

Fault attacks are active attacks, which need an adversary to induce errors into the target device, using some tampering means. This tampering can be accomplished in several ways, as extensively discussed in literature and illustrated in Fig. 2.1. In general tampering means or fault injection techniques are classified in two broad categories, i.e., global and local. Global fault injection are, in general, low-cost techniques which create disturbances on global parameters like voltage, clock, etc. The resultant faults are more or less random in nature and the adversary might need several injection, to find required faults. On the other hand, local techniques are more precise in terms of fault location and model. However, this precision needs expensive equipment.

The kind of fault injected can be defined as fault model. The fault model has two important parameters, i.e., location and impact. Location means the spatial and

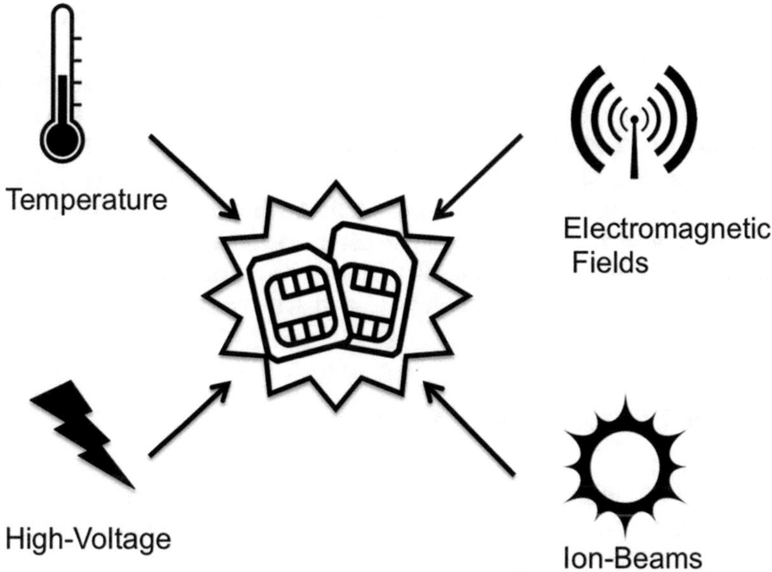

Fig. 2.1 An illustration of fault injection techniques

temporal location of fault injection during the execution of target algorithm. Depending on the type and precision of the technique, location can be at the level of *bit*, *variable* or *random*. Coming to the impact of fault, it is the affect on the target data. Commonly known fault injection impacts on target data can cause *stuck-at*, *bit-flip*, *random-byte*, or *uniformly distributed random value*. In the rest of the section, we summarize the most common techniques for injecting faults, highlighting for each of them, the main characteristics, the complexity, the cost, and the erroneous behavior introduced in the target device.

2.2.1 Fault Injection Through Power Supply

Embedded systems are often either battery operated or connected with an external power supply (latter, for instance, the case of smart cards). In this context, a natural and very inexpensive way to induce a malfunctioning is to alter the power supply coming from an external source. This alteration or disturbance can be performed in two different ways: *underfeeding* or *voltage glitch*. The typical effects caused by alteration of power supply are setup-time violation: flip-flops are triggered before the input signals reach a stable and correct value. Such fault techniques can be used to skip the execution of an instruction in the microprocessor code. The temporal precision of the fault injection depends on the accuracy of the voltage drop in duration and its synchronization with the target device. By underfeeding the target for a prolonged period, the adversary is able to insert transient faults with single-bit faults appearing first and increasing in multiplicity as the feeding voltage is further lowered. This requires only basic skill and can be easily achieved in practice without leaving evidence of tampering on the device. Alterations of power supply was exploited by Barenghi et al. [5] to attack software implementations of the AES and the RSA algorithms. The authors demonstrated that an embedded processor, in that case an ARM9, can be successfully attacked using very inexpensive equipment. Several other works demonstrated the feasibility of attacking the AES algorithm with fault induced by underfeeding. A possible example is the work of Selmane et al. [6]. The drawback of this technique is the time precision: the adversary can not control the exact time in which the fault happens. As a results, he must be capable of selecting the right faults and discard the ones which cannot be used during attacks.

2.2.2 Fault Injection Through Clock

Fault induced with clock are similar to fault induced with power supply. The typical target of these attacks are devices, such as smartcards, which use an external clock signal. The adversary can supply these devices with an altered clock signal, which contains, for instance, a clock pulse which is much shorter than what is expected in the normal clock. Pulses generated in this way are called clock glitches, are much shorter than the deviations normally tolerated by the smart cards [7, 8], and they can cause a setup-time violation or the skipping of instructions during the execution of a program [9]. The errors are transient and the device does not incur any damage; thus it is possible to induce faults at will without leaving any evidence of tampering. Overclocking the target for a prolonged period can also be used to inject transient faults. Much like underfeeding, overclocking leads to single-bit flips initially and higher multiplicity of faults at higher frequencies. This type of attack can be carried out with relatively low cost equipment.

2.2.3 *Fault Injection Through Temperature*

Electronic circuit are manufactured to work only on specific operating conditions identified by an upper and a lower temperature thresholds. Outside this range, there is no guarantee that the circuit continues to work as expected. Possible effects can be the random modification of the content of the memory cells or a limited functionality of the device. An adversary can induce faults into the device by exposing it to temperatures outside this range or by stressing it in order to increase its temperature. An example of this approach is the one proposed by Govindavajhala and Appel [10], where the chip was heated by executing a large number of load and store operations. The authors report a thermal fault induction attack against the DRAM memory chips of a common desktop PC. Using a 50 W light bulb and controlling its distance from the target, the authors reported around ten flipped bits per 32-bit word at 100 °C. Since the precision of the heating element is coarse grain, the controllability of faults is limited and impact is global. Moreover, excessive heating can cause permanent damage to the target.

2.2.4 *Fault Injection Through Light*

Optical attacks are a semi-invasive fault injection attack because they require the decapsulation of the target device, which is then hit with a light pulse or high-intensity laser. The light pulse or laser can be directed to the front or to the back side of the chip, depending on the type of the attack and the difficulty involved in each approach. In fact, with modern technologies, it can be difficult to reach the target cell from the front side of the chip, due to several metal layers of the chip itself. In order to obtain very precisely focused light pulses, the light emitted from a camera flash is concentrated with the aid of a precision optical microscope by applying it to the eyepiece after the device under attack has been carefully placed on the slide holder. In order to avoid over-irradiation of the device, which might lead to permanent damage to the circuit, care must be taken in selecting an appropriate magnification level for the microscope lens. The accuracy with light pulse is limited as the pulse gets scattered. Laser provides higher accuracy. It can also be used to attack through the back-substrate of the chip using near-infrared wavelengths. Optically induced fault require expensive equipment to be carried out. High-energy radiations like UV lamps can also create permanent faults. Nevertheless, they can be very precise both in the target, as it was demonstrated by Skorobogatov [11], and in time. This precision allow to change values at the granularity of a single RAM cell or even at of a register [12]. Light-based faults need medium to high expertise depending on the equipment. Laser is also capable of damaging the chip with over-radiation. The faults can be precise in time and space and also multiple faults can be injected using laser.

2.2.5 Fault Injection Through Electromagnetic Fields

Electromagnetic pulses can cause the change in the memory content or the malfunctioning of the device. This is due to the so called Eddy currents, which are created using an active coil [13]. Electromagnetic pulses can induce a fault which is extremely localized and precise (up to the level of a single bit), while the equipment needed to carry out this attack can be relatively cheap. Furthermore, the attack can be carried out without depackaging the chip. However, the adversary is required to know the details of the layout of the chip in order to identify the precise point of attack. The EM pulse can either be injected over the power trail of the chip, uniformly affecting the whole attacked device, or a smaller EM coil can be used to induct an additional current on a specific part of the circuit. The idea of this type of fault injection has been introduced by Quisquater and Samyde [13]. The authors demonstrated that it is possible to alter the computations of a cryptographic algorithm using an Electromagnetic probe and a camera flash (used to induce high voltage into the coil of the probe). This technique does not work efficiently with chips that employ grounded metal packaging (usually for heat sinking purposes), that also act as an EM shield, which needs the adversary to perform decapsulation (Table 2.1).

2.2.6 Fault Injection Through Focused Ion Beams

Focused Ion Beams (FIB) are a very expensive way to inject fault into the device, however they allows the attacker to arbitrarily modify the structure of a circuit. The adversary can cut existing wires, add connections, and operated through different layers. The capability of these tools are demonstrated in Torrance et al. [14], where the authors showed the reconstruction of a bus without damaging the contents of the memory. FIB equipment is expensive and needs high technical expertise, but the precision is extremely high. Current FIBs are able to operate with precision up to 2.5 nm, i.e., less than a tenth of the gate width of the smallest transistor that can currently be etched.

Table 2.1 Comparison of fault injection techniques

Technique	Type	Target			Precision		Equipment	Expertise
		Decapsulation	Invasive	Design details	Time	Space	Costs	
Clock glitch	Global	No	Non-invasive	Required	High	Low	Low	Moderate
Voltage glitch	Global	No	Non-invasive	Partial needed	Moderate	Low	Low	Moderate
Underfeeding	Global	No	Non-invasive	No	None	High	Low	Low
Overclocking	Global	No	Non-invasive	No	None	High	Low	Low
Temperature	Global	No	Semi-invasive	Required	None	Low	Low	Low
Global EM pulse	Global	No	Semi-invasive	Partial needed	Moderate	Low	Low	Moderate
Local EM pulse	Local	Sometimes	Semi-invasive	Required	Moderate	Moderate	Moderate	Moderate
Light pulse	Local	Yes	Semi-invasive	Required	Moderate	Moderate	Moderate	Moderate
Laser beam	Local	Yes	Semi-invasive	Required	High	High	High	High
Light radiation	Local	Yes	Invasive	No	Low	Low	Low	Moderate
Focused ion beam	Local	Yes	Invasive	Required	Complete	Complete	Very high	Very high

2.2.7 Comparison of Fault Injection Techniques

The previously introduced fault injection techniques have several parameters to define its application. The most important parameters are compared in Table 2.2.

2.3 Fault Attacks

Faults attacks have gained popularity as a serious threat to embedded systems over the last few years. Attacks can target a specific algorithm or generically modify the program flow to attacker's advantage. In the following, we refer the classification of attacks and the organization proposed by Karaklajić et al. [15]. In particular, three distinct classes of fault attacks are identified for embedded system.

2.3.1 Algorithm Specific Attacks

Fault attacks can be designed to exploit specific weaknesses of the target algorithm which are introduced by the injection of a fault. Several attacks targeting a large number of algorithms were presented in the past, the most common being the attacks against AES, RSA, and ECC.

Bloemer et al. in [16] proposed an attack on AES which exploit the change of a single bit after the first key addition. However, this attack can successfully recover a complete key only when the adversary has the possibility to inject a fault at a very precise timing and at a very specific position.

The security of asymmetric cryptosystems relies on problems which are mathematically hard to be solved. Fault attacks can be designed to weaken the problems and thus weaken the security of the algorithm based on that. A common target for such attacks are public-key cryptography algorithm, in particular RSA and ECC, as they are widely used for authentication, digital signature, and key exchange. RSA is based on exponentiation using a square and multiply (S&M) routine, while ECC is based on point-scalar multiplication using a double and add (D&A) routine. Both (S&M) and (D&A) have similar structure where the set of executed routine depends on the value of the processed bit of the secret.

Proposed attacks to these cryptosystems requires the attacker to change the base point of an ECC. As a result, the scalar point-multiplication will be moved to a weaker curve. The use of weak curve will make the problem of solving the discrete-logarithm problem of ECC manageable, and thus will lead to the recover of the secret [17]. The same attack can be carried out if the attacker manage to supply wrong parameters for the curve [17]. Other attacks proposed in the past showed that faults can be exploited to control few bits of the secret nonce in DSA and, which ultimately allows to recover the whole key [18]. Pairing algorithm are also vulnerable

Table 2.2 Comparison of fault injection mechanisms

Mechanism		Cost	Controllability	Trigger	Type	Repeatability	Injection time	Risk of damage	Runtime injection
Simulator	Static analisys								No
	Execution based	Med.	High	Yes	App. [lim.]	High	Med.	No	No
	Trace-based				Sys. [lim.]				Yes
	Transistor level								No
Software sim.	Compile time	Low	Low	Yes	App.	High	Low	No	No
	Runtime				OS				Yes
Low-level VM sim.		Med	High	Yes	OS Sys [lim].	Med.	High	No	Yes
Emulation		High	Med.	Med.	App. [lim.]	Med.	High	Yes	Yes
					Sys.				
Hardware		High	Med.	Med.	Sys.	Med.	High	Yes	Yes

to faults [19]: it was demonstrated that by modifying the loop parameter of a pairing algorithm, the secret point can be recovered.

2.3.2 Differential Fault Analysis

Differential fault analysis (DFA) is one of the most common exploits for cryptographic algorithms. The main idea behind DFA is the following: faults are injected in a device to alter the computation of the target algorithm. When a fault injection is successful, it is reflected on the output ciphertext. The attacker also computes the correct ciphertext for the same inputs. After collecting several correct/faulty ciphertext pairs, differential cryptanalysis techniques are applied to discard key candidates. With precise faults injection, the attacker can reduce the key candidates to a unique solution. DFA was introduced as a threat to symmetric block ciphers by Biham [20] and then extended to other symmetric primitives [21, 22]. The attack presented by Giraud [23] targets the state during the last round, and assumes that the fault alters only a single bit of the state, prior to the last SubBytes operation. The attack is thus applicable only to the last AES round that does not include the MixColumns operation, where the a single byte difference in the state will not spread to other bytes. Attacks proposed in the past also target a single bit or a single byte in the key-expansion routine of the target algorithm in order to recover the whole secret key [24].

DFA has also been reported on public-key algorithms like RSA [25]. Bellcore attack [26] attempts to factor the modulus n by injecting faults in exponentiation phase using the Chinese remainder theorem. DFA was further extended to fault sensitivity analysis (FSA [27]), which exploits the physical characteristics of faults like timing instead of faulty outputs.

2.3.3 Tampering with the Program Flow

Faults can also be injected to change the flow of an executed software code [28–31]. A fault in program counter is an obvious way to modify the program flow. For code implementing cryptographic algorithms, change in the program flow can be a security threat. Often an instruction skip lead to a wrong computation of critical portion of the algorithm, significantly weakening it. A notable example of this involves the exponentiation algorithm [28]. Similarly, fault on loop counter or branch selection were demonstrated to leak secret point of pairing schemes [30] or to reduce the number of encryption round in symmetric key algorithms [20], thus enabling classical cryptanalysis. Often fault attack countermeasures are based on redundancy with sanity check. Some efficient attacks simply try to skip the sanity check to bypass a deployed countermeasure [32]. This approach can be distinguished in two main categories: during compile time or during runtime. To inject faults at compile time, the program instruction must be modified before the program image is loaded and executed. Rather than injecting faults into the hardware of the target system, this method injects errors into the source code or assembly code of the target program to emulate the effect of hardware, software, and transient faults. During runtime, a mechanism is needed to trigger fault injection. Commonly used triggering mechanisms include, timeouts, exception/traps and code insertions. Although this approach is flexible, it has its shortcomings: first of all, it cannot inject faults into locations that are inaccessible to software. Second, the poor time-resolution of the approach may cause fidelity problems. For long latency faults, such as memory faults, the low time-resolution may not be a problem. For short latency faults, such as bus and CPU faults, the approach may fail to capture certain error behavior, like propagation.

2.4 Fault Injection Simulators and Their Applicability to Fault Attacks

In simulation-based fault injection, the target system as well as the possible hardware faults are modeled and simulated by a software program, usually called fault simulator. The fault simulation is performed by modifying either the hardware model or the software state of the target system. This means that the system could behave as if there was a hardware fault [13]. There are two categories of fault injection: runtime fault injection and compile-time fault injection. In the former, faults are injected during the simulation or the execution of the model. In the latter, faults are injected at compile time in the target hardware model or in the software executed by the target system. The advantage of the simulation-based fault injection techniques is that there is no risk to damage the system in use. In addition, they are cheaper in terms of time and efforts than the hardware techniques. They also have a higher controllability and observability of the system behavior in the presence of faults. Nevertheless, simulation-based fault injection techniques may lack in the accuracy of the fault model and the system model. In addition, they have a poor time-resolution, which may cause fidelity problems. Software fault injection is a special case of simulation-based fault injection where the target system is a large microprocessor-based machine that may include caches, memories, and devices, running a complex software. This technique is able to target applications and operating systems, which is not easy to do with the hardware fault injection.

Fault-injection simulators are attractive because they do not require expensive hardware. Moreover, they can be used to support all system abstraction levels, as applications and operative systems, which is difficult at hardware level. The controllability of fault-injection simulators is very high: given sufficient detail in the model, it is possible to modify any signal value in any desired way, with the results of the fault-injection easily observable regardless of the location of the modified signal within the model. The main goal of an early analysis of the resistance against fault attacks is to allow designers to easily identify the weakest point of their design, and to protect it with appropriate countermeasures. Although this approach is flexible, it has some shortcomings:

- Large development efforts are required, as they involve the construction of the simulation model of the system under analysis, including a detailed model of the processor in use. This increase the cost of using simulation-based fault-injection tools.
- Not all the fault attacks previously discussed can be simulated in the simulation model.
- The fidelity of the model strongly depends on the accuracy of the models used.
- High time consuming, due to the length of the experiment.

Some attacks, in particular setup-time violations, can be reliably simulated using state of the art EDA commodities. For some others, instead, it is impossible to have a complete simulation. It is however possible to model the type of error which will be induced into the device, and simulate the behavior of a device when a similar type of error occurs with cycle accurate or with behavioral simulators. The strategy usually adopted by these injection frameworks is to evaluate the effects, that the injected faults have on the final result of the computation. Designer then attempts to mount an attacks using the simulated data and can determine if the amount of information which will be available to the attacker will be sufficient to successfully extract secret information. In the rest of this section we revise known tools and approaches used in the past for injecting and simulating faults at different level of abstraction and we discuss their suitability for evaluating the resistance against fault attacks.

2.4.1 Weaknesses Identification with Static Analysis

Identification of portions of the circuit sensitive to fault attacks can be achieved using static timing analysis. Static timing analysis produces a very detailed timing characterization of the paths inside their design, highlighting the critical path and all the other paths which are very close to the critical one. Barenghi et al. [33] proposed to extract the worst-case delays associated with the input connections of the state and key registers. Static analysis was carried out with Synposys PrimePower, using as input the placed and routed netlist and the parasitics of the connections. The authors compared the ranking of sensitivity to attacks, obtained using static analysis, with the fault attacks mounted on a real device. Obtained results demonstrated that static timing analysis provides an effective way to estimate the worst case timings for the input lines of the state registers and pinpoint which ones are more likely to be vulnerable to setup-time violation attacks.

2.4.2 High-Level Simulation with Complex Fault Models

High-level simulators are system simulator which simulate the behavior of a device with the precision of a clock cycle. They can be execution based, when the benchmark is directly executed, or trace based, when the simulation is carried out using a trace of execution previously generated.

2.4.2.1 Fault Injection in Execution-Based Simulators

In these kind of simulators, a module for fault injection is integrated in the target design. The fault injection module can be integrated as dedicated module called *saboteur*. It is inactive during normal operation and can alter value or timing characteristics when active. Saboteurs can be inserted in series or parallel to the target design. Serial insertion, in its simplest form, consists of breaking up the signal path between a driver (output) and its corresponding receiver (input) and placing a saboteur in between. In its more complex form, it is possible to break up the signal paths between a set of drivers and its corresponding set of receivers and insert a saboteur. For parallel insertion, a saboteur is simply added as an additional driver for a resolved signal. The other approach of fault injection is using *mutants* which are inserted by modifying parts of the target circuit components. Those two approaches present the advantage of supporting all system abstraction levels: electrical, logical, functional, and architectural. Such approaches allow full reproduction of: *single-bit flips, selected bit alterations, data corruptions, circuit rewiring, clock alteration* and *instruction swaps* effects. However, theory require large development effort and cannot support fully randomisation and real-time features.

Existing available tools are: MEFISTO-C [34], VERIFY [35], HEARTLESS [36], GSTF [37], FTI [38], Xception [39], FERRARI [40], SAFE [41], DOCTOR [42]. A detailed overview can be found in [43].

Selected Simulators

Xception [39] is a software implemented fault injection tool for dependability analysis. It provides an automated test suite that helps in injecting realistic faults. It injects faults without any intrusion on the target system. No software traps are inserted and hence program can be executed in normal speed. It uses the advanced debugging and performance monitoring features that exist in processors to inject realistic faults by software, and to monitor the activation of the faults in order to observe in detail their impacts on the behavior of the system [39]. Xception is a flexible and low-costly tool that could be used in a wide range of processors and machines (parallel and real-time systems). In addition, it enables the definition of a general and precise processor fault model with a large range of fault triggers and oriented to the internal processor functional units.

2.4.2.2 Software Faults Emulation Tools

A few tools do exist to emulate the occurrence of faults in distributed applications. One of those tools is DOCTOR [42] (integrateD sOftware fault injeCTiOn enviRonment), that allows to inject faults in real-time systems. It supports faults in processor, memory and communication. The tool can inject permanent, transient, or intermittent faults. The fault scenarios that can be designed uses probabilistic model. While this suits small quantitative tests, repeatable fault injection capabilities are required for more complex fault scenarios.

SAFE [41] fault injection tool allows to automatically generate and execute fault injection tests. SAFE injects or detects software faults in C and C++ software, in order to force a software component failure, and to evaluate the robustness of the system as a whole. Injected faults are designed to realistically reproduce the real defects that hampers software systems, including issues affecting data initialization, control flow, and algorithms. Testing team can easily know how vulnerable the software is and fix it. The SAFE tool lets users customize which faults are injected.

2.4.2.3 Fault Injection in Trace-Based Simulators

An example of fault injection tools exploiting trace-based simulations is the one of Miele [44]. The tool analyzes the system-level dependability of embedded systems. The workflow is organized in three main phases: preliminary characterization of the system, setup of the experimental campaign, and execution of experimental campaign followed by results' post-processing. The designer specifies monitoring and classification actions at application and architecture levels. Debug-like mechanism allow to analyze the propagation of the errors in various functionalities of the executed application. The proposed approach is extremely suitable to reproduce the effects in simulation of *single-bit flips, selected bit alterations, data corruptions* and *instruction swaps*. Ferrari [40] (Fault-and-Error Automatic Real-Time Injection), developed at the University of Texas at Austin, uses software traps to inject CPU, memory, and bus faults. Ferrari consists of four components: the initializer and activator, the user information, the fault-and-error injector, and the data collector and analyzer. The fault-and-error injector uses software trap and trap handling routines. Software traps are triggered either by the program counter when it points to the desired program locations or by a timer. When the traps are triggered, the trap handling routines inject faults at the specific fault locations, typically by changing the content of selected registers or memory locations to emulate actual data corruptions. The faults injected can be those permanent or transient faults that result in an address line error, a data line error, and a condition bit error.

Jaca is a fault injection tool that is able to inject fault in object-oriented systems and can be adapted to any Java application without the need of its source code, but only few information about the application like the classes, methods, and attributes names [45]. Jaca has a graphical interface that permits the user to indicate the applications parameters under test in order to execute the fault injection [45]. Most of the fault injection tools are able to handle the injection of faults at low level of the software. Jaca differs from the other tools in the fact that it can perform both low-level fault injection, affecting Assembly language element (CPU registers, buses, etc.), and high-level fault injection affecting the attributes and methods of objects in a Java program.

The main advantage of using a trace-based simulator is the possibility of altering specific parts of the system without the need of altering the main structure of the system.

2.4.2.4 Software-Based Simulators

Software fault injection is a special case of simulation-based fault injection where the target system is a large microprocessor-based machine that may include caches, memories, and devices, running a complex software. This technique is able to target applications and operating systems, which is not easy to do with the hardware fault injection.

Selected Simulators

LFI is a tool to make fault injection-based testing more efficient and accessible to developers and testers [46]. LFI injects faults at the boundary between shared libraries and target programs, which permits to verify if the programs are handling the failures exposed by the libraries correctly or not. More in detail, LFI permits to automatically identify the errors exposed by shared libraries, find potentially buggy error recovery code in program binaries, and produce corresponding injection scenarios. Fault injection was rarely used in software development. LFI was developed in response to this. It permits to reduce the dependence on human labor and correct documentation, because it automatically profiles fault behaviors of libraries via static analysis of their binaries. The tool aims to provide testers an easy, fast, and comprehensive method to see how much the program is robust to face failures exposed between shared libraries and the tested programs [46].

Byteman [47] is a byte code injection tool developed to support Java code testing using fault injection technique. It is also very useful for troubleshooting and tracing Java program execution. Byteman provides a functions library which helps generating simple error conditions to complex error flows. Almost any Java code can be injected into the application in scope at the injection point. POJO (plain old java object) can be plugged in to replace built-in functions. Byteman works by modifying the bytecode of the application classes dynamically at runtime.

2.4.3 Low-Level Virtual Machine Simulation

Fault injection tools based on virtual machine can be a good solution. First, because they permit simulating the computer without having the real hardware system.

Moreover, they target hardware faults on the software level, and they allow observing complex computer-based systems, with operating system and user applications.

Virtualization is the technology permitting to create a virtual machine (VM) that behaves like a real physical computer with an Operating System (OS). It has an enormous effect in todays IT world since it ensures efficient and flexible performance, and permits cost saving from sharing the same physical hardware. The virtual machine where the software is running is called a guest machine, and the real machine in which the virtualization takes place is called the host machine. The words host and guest are used to make difference between the software that runs on the virtual machine and the software that runs on the physical machine.

Selected Simulators

LLVM (Low-Level Virtual Machine) is a compiler framework designed to support transparent, life-long program analysis, and transformation for arbitrary programs, by providing high-level information to compiler transformations at compile time, link-time, runtime, and in idle-time between runs [48, 49].

LLVM uses the LLVM Intermediate Representation (IR) as a form to represent code in the compiler. It symbolizes the most important aspect of LLVM, because it is designed to host mid-level analysis and transformations found in the optimizer section of the compiler. The LLVM IR is independent from the source language and the target machine. It is easy for a front end to generate, and expressive enough to permit important optimizations to be performed for real targets.

QEMU [50] is a versatile emulation platform with support for numerous target architectures like x86, ARM, MIPS and allowing to run a variety of unmodified guest operating systems. In [51], BitVaSim is proposed as a fault injection simulator on QEMU platform, for targets like PowerPC and ARM with built-in test software framework. BitVaSim can inject faults in any process, even pre-compiled software and allows a good degree of user configuration for fault injection. It can also be used for hardware targets in a virtual machine. In addition, unmodified operating systems and applications, especially the Built-In Test system can run on the prototype without intrusion. The described technique provides complete control over the target environment with fault injection process monitor and efficient feedback.

FAUMachine is a virtual machine that permits to install a full operating systems and run them as if they are independent computers. FAUMachine is similar in many aspect to standard virtual machines like QEMU [50] or VirtualBox [52]. The property that distinguishes FAUMachine from the other virtual machines is its ability to support fault injection capabilities for experimentation. FAUMachine supports the following fault types [53]:

- Memory Cells: such as transient bit flips, permanent struck-at faults, and permanent coupling faults.
- Disk, CD/DVD drive: such as transient or permanent block faults, and transient or permanent whole disk faults.
- Network: such as transient, intermittent, and permanent send or receive faults.

FAUMachine does not permit injecting faults in the CPU registers yet. Bit flips could be easy to implement. Stuck-at faults is also possible but it is much more complex since FAUMachine uses just-in-time compiling. In FAUMachine, the injection of fault could be done online via GUI, or defined (type, location, time, and duration of fault) via VHDL scripts. Compared to existing fault injection tools, FAUMachine is able to inject faults and observe the whole operating system or application software. Using the virtualization, this tool provides a high simulation speed for both complex hardware and software systems [53]. FAUMachine also supports automated tests, including the specification of faults to be injected.

2.4.4 Transistor Level Simulation

As previously discussed, setup-time violation can be induced by underfeeding the device. This attack can be completely simulated using SPICE level simulators, as proposed by Barenghi et al. [33]. The authors evaluated if transistor level simulator is capable of correctly predicting the fault patterns which were measured on a real device. The simulation was carried out using Synopsys Nanosim, a fast SPICE simulator, using the netlist and the parasitics generated by Cadence Encounter after place and route. The device was simulated for different voltages, ranging from 0.3 to 0.5 V. The simulation generated a number of faulty ciphertexts reasonably close to the one observed in the experiments, allowing to speculate that Nanosim is capable of predicting the setup-time violations measured in practice.

2.4.5 Emulation

Emulation-based fault injection has been introduced as a better solution for reducing the execution time compared to simulation-based fault injection. It is often based on the use of Field Programmable Gate Arrays (FPGAs) for speeding up fault simulation and exploits FPGAs for effective circuit emulation. This technique can allow the designer to study the actual behavior of the circuit in the application environment, taking into account real-time interactions. However, when an emulator is used, the initial VHDL description must be synthesizable.

Fault injection can be performed in hardware emulation models through compile time reconfiguration and runtime reconfiguration. Here reconfiguration refers to the process of adding hardware structures to the model which are necessary to perform the experiments. In compile-time reconfiguration, these hardware structures are added by the instrumentation of HDL models. The main disadvantage of compile-time reconfiguration is that the circuit must be resynthetised for each reconfiguration, which can impose a severe overhead on the time it takes to conduct a fault injection campaign.

2.5 Conclusions

Faults attacks are a powerful tool in the hand of adversaries, and they can have serious impacts on the security of embedded systems. Currently, most of the evaluation against fault attacks is done post-fabrication. However, it is important for designers to know the sensitive fault targets and possibly fix it at the design time. With this objective in mind, we summarize the most common fault attacks, the most frequently used fault injection techniques and the most used approach for fault simulations which could be used to evaluate the robustness of cryptographic circuits at design time.

Table 2.2 classifies the different fault injection methods. The approaches presented have different level of impact. Static timing analysis [33] provides an effective way for the designer to predict circuit paths which are likely to experience setup-time violations upon an attack, but it does not provide the possibility of simulating a fault.

Saboters and *mutant* fault injection approaches allow to properly simulate fault injection effects such as *single bit flips, selected bit alterations, data corruptions, circuit rewiring, clock alteration*, and *instruction swaps*. Furthermore, they provide full control of both fault models and injection mechanisms, together with maximum amount of observability and controllability. Essentially, given sufficient detail in the model, any signal value can be corrupted in any desired way, with the results of the corruption easily observable regardless of the location of the corrupted signal within the model. This flexibility allows any potential failure model to be accurately modeled. These methods are able to model both transient and permanent faults, and allow modeling of timing-related faults since the amount of simulation time required to inject the fault is minimal. The main drawback of those two approaches is given by the fact that only a predetermined set of faults can be injected, and new changes cannot be applied at runtime.

The main advantage of using a trace-based simulator as in [44] is given by the possibility of changing at runtime the execution traces, without a structural modification of specific components of the architecture or saboteurs. In this way, fault effects generation is easier and less time consuming. Moreover, it allows for time-specific fault attacks, since it is a cycle accurate based simulator. As main disadvantage, not all the current existing fault effects such as rewiring and clock delays can be effectively simulated, thus prohibiting an exhaustive analysis.

Virtual Machines-based tools can be a good solution for injecting faults. First, because they permit simulating the computer without having the real hardware system. Moreover, they target hardware faults on the software level, and they allow observing complex computer-based systems, with operating system and user applications. However, at the state of the art, they do not support yet fault models defined at software level, such as an instruction or a variable used in place of another.

Finally, for certain fault injection techniques, a complete simulation at SPICE level is possible. The drawback of this approach is the time required for the simulation, which can be prohibitive.

References

1. Kocher P, Jaffe J, Jun B. Differential power analysis. In: Advances in CryptologyCRYPTO99. Springer; 1999. p. 388–97.
2. Kocher PC. Timing attacks on implementations of Diffie-Hellman, RSA, DSS, and other systems. In: Advances in CryptologyCRYPTO96. Springer; 1996. p. 104–13.
3. Rohatgi P. Electromagnetic attacks and countermeasures. In: Cryptographic engineering. Springer; 2009. p. 407–30.
4. Schlösser A, Nedospasov D, Krämer J, Orlic S, Seifert J-P. Simple photonic emission analysis of AES. In: Cryptographic hardware and embedded systems—CHES 2012. Springer; 2012. p. 41–57.
5. Barenghi A. Bertoni GM, Breveglieri L, Pelosi G. A fault induction technique based on voltage underfeeding with application to attacks against AES and RSA. J Syst Softw. 2013;86(7):1864–78.
6. Selmane N, Guilley S, Danger J-L. Practical setup time violation attacks on AES. In: Seventh European dependable computing conference, EDCC-7 2008, Kaunas, Lithuania, 7–9 May 2008, IEEE Computer Society; 2008. p. 91–6.
7. Otto M. Fault attacks and countermeasures. PhD thesis, Universit at Paderborn; 2005.
8. International organization for standardization. ISO/IEC 7816-3: electronic signals and transmission protocols. 2002. http://www.iso.ch.
9. Balasch J, Gierlichs B, Verbauwhede I. An in-depth and black-box characterization of the effects of clock glitches on 8-bit MCUs. In: Breveglieri L, Guilley S, Koren I, Naccache D, Takahashi J, editors. 2011 workshop on fault diagnosis and tolerance in cryptography, FDTC 2011, Tokyo, Japan, September 29, 2011. IEEE; 2011. p. 105–14
10. Govindavajhala S, Appel AW. Using memory errors to attack a virtual machine. In: Proceedings of the 2003 IEEE symposium on security and privacy, SP '03. Washington, DC, USA: IEEE Computer Society; 2003. p. 154.
11. Skorobogatov S. Optical fault masking attacks. In: Breveglieri L, Joye M, Koren I, Naccache D, Verbauwhede I, editors. 2010 workshop on fault diagnosis and tolerance in cryptography, FDTC 2010, Santa Barbara, California, USA, 21 August 2010. IEEE Computer Society; 2010. p. 23–9
12. Barenghi A, Breveglieri L, Koren I, Naccache D. Fault injection attacks on cryptographic devices: theory, practice, and countermeasures. Proc IEEE. 2012;100(11):3056–76.
13. J-J Quisquater, D Samyde. Eddy current for magnetic analysis with active sensor. In: Esmart 2002, Nice, France, 9 2002.
14. Torrance R, James D. The state-of-the-art in IC reverse engineering. In: Cryptographic hardware and embedded systems-CHES 2009. Springer; 2009. p. 363–81.
15. Karaklajic D, Schmidt J-M, Verbauwhede I. Hardware designer's guide to fault attacks. IEEE Trans VLSI Syst. 2013;21(12):2295–306.
16. Blömer J, Seifert J-P. Fault based cryptanalysis of the advanced encryption standard (AES). In: Wright RN, editor. Financial cryptography, 7th international conference, FC 2003, Guadeloupe, French West Indies, January 27–30, 2003, revised papers. Lecture notes in computer science, vol. 2742. Springer; 2003. p. 162–81.
17. Ciet M, Joye M. Elliptic curve cryptosystems in the presence of permanent and transient faults. Des Codes Crypt. 2005;36(1):33–43.
18. Naccache D, Nguyen PQ, Tunstall M, Whelan C. Experimenting with faults, lattices and the DSA. In: Vaudenay S, editor. Public key cryptography—PKC 2005, proceedings of the 8th international workshop on theory and practice in public key cryptography, Les Diablerets, Switzerland, January 23–26, 2005. Lecture notes in computer science, vol. 3386, Springer; 2005. p. 16–28.
19. Duursma IM, Lee H-S. Tate pairing implementation for hyperelliptic curves $y^2 = x^p - x + d$. In: Laih C-S, editor. ASIACRYPT 2003, proceedings of the 9th international conference on the theory and application of cryptology and information security: advances in cryptology,

Taipei, Taiwan, November 30–December 4, 2003. Lecture notes in computer science, vol. 2894. Springer; 2003. p. 111–23.
20. Biham E, Shamir A. Differential fault analysis of secret key cryptosystems. In: Kaliski BS Jr., editor. CRYPTO '97, proceedings of the 17th annual international cryptology conference on advances in cryptology, Santa Barbara, California, USA, August 17–21, 1997. Lecture notes in computer science, vol. 1294. Springer; 1997. p. 513–25.
21. Tunstall M, Mukhopadhyay D, Ali S. Differential fault analysis of the advanced encryption standard using a single fault. In: Ardagna CA, Zhou J, editors. Proceedings of the 5th IFIP WG 11.2 international workshop on information security theory and practice. Security and privacy of mobile devices in wireless communication, WISTP 2011, Heraklion, Crete, Greece, June 1–3, 2011. Lecture notes in computer science, vol. 6633. Springer; 2011. p. 224–33.
22. Rivain M. Differential fault analysis of DES. In: Joye M, Tunstall M, editors. Fault analysis in cryptography. Information security and cryptography. Springer; 2012. p. 37–54
23. Giraud C. Dfa on aes. In: Advanced encryption standard—AES, 4th International conference, AES 2004. Springer; 2003. p. 27–41.
24. Biehl I, Meyer B, Müller V. Differential fault attacks on elliptic curve cryptosystems. In: Bellare M, editor. CRYPTO 2000, proceedings of the 20th annual international cryptology conference on advances in cryptology, Santa Barbara, California, USA, August 20–24, 2000. Lecture notes in computer science, vol. 1880. Springer; 2000. p. 131–146.
25. Berzati A, Canovas C, Goubin L. Perturbating RSA public keys: an improved attack. In: Oswald E, Rohatgi P, editors. CHES 2008, proceedings of the 10th international workshop on cryptographic hardware and embedded systems, Washington, D.C., USA, August 10–13, 2008. Lecture notes in computer science, vol. 5154. Springer; 2008. p. 380–395.
26. Boneh D, DeMillo RA, Lipton RJ. On the importance of checking cryptographic protocols for faults (extended abstract). In: Fumy W, editor. Advances in cryptology—proceedings of the EUROCRYPT '97, international conference on the theory and application of cryptographic techniques, Konstanz, Germany, May 11–15, 1997. Lecture notes in computer science, vol. 1233. Springer; 1997. p. 37–51.
27. Li Y, Sakiyama K, Gomisawa S, Fukunaga T, Takahashi J, Ohta K. Fault sensitivity analysis. In: Mangard S, Standaert F-X, editors. CHES 2010, proceedings of the 12th international workshop on cryptographic hardware and embedded systems, Santa Barbara, CA, USA, August 17–20, 2010. Lecture notes in computer science, vol. 6225. Springer; 2010. p. 320–34.
28. Schmidt J-M, Herbst C. A practical fault attack on square and multiply. In: Breveglieri et al. [56], p. 53–8.
29. Schmidt J-M, Medwed M. A fault attack on ECDSA. In: Breveglieri L, Koren I, Naccache D, Oswald E, Seifert J-P, editors. Sixth international workshop on fault diagnosis and tolerance in cryptography, FDTC 2009, Lausanne, Switzerland, 6 September 2009. IEEE Computer Society; 2009. p. 93–9.
30. Page D, Vercauteren F. A fault attack on pairing-based cryptography. IEEE Trans Comput. 2006;55(9):1075–80.
31. Schmidt J-M, Medwed M. Fault attacks on the montgomery powering ladder. In: Rhee KH, Nyang DH, editors. Information security and cryptology—ICISC 2010: 13th international conference, Seoul, Korea, December 1–3, 2010, revised selected papers. Lecture notes in computer science, vol. 6829. Springer; 2010. p. 396–406.
32. Kim CH, Shin JH, Quisquater J-J, Lee PJ. Safe-error attack on SPA-FA resistant exponentiations using a HW modular multiplier. In: Nam K-H, Rhee G, editors. ICISC 2007, proceedings of the 10th international conference on information security and cryptology, Seoul, Korea, November 29–30, 2007. Lecture notes in computer science, vol. 4817. Springer; 2007. p. 273–81.
33. Barenghi A, Hocquet C, Bol D, Standaert F-X, Regazzoni F, Koren I. A combined design-time/test-time study of the vulnerability of sub-threshold devices to low voltage fault attacks. IEEE Trans Emerg Top Comput. 2014;2(2):107–18.
34. Folkesson P, Svensson S, Karlsson J. A comparison of simulation based and scan chain implemented fault injection. In: Digest of papers: FTCS-28, the twenty-eigth annual international symposium on fault-tolerant computing, Munich, Germany, June 23–25, 1998. IEEE Computer Society; 1998. p. 284–93.

35. Sieh V, Tschäche O, Balbach F. Verify: evaluation of reliability using vhdl-models with embedded fault descriptions. In: FTCS. IEEE Computer Society; 1997. p. 32–6.
36. Rousselle C, Pflanz M, Behling A, Mohaupt T, Vierhaus HT. A register-transfer-level fault simulator for permanent and transient faults in embedded processors. In: DATE; 2001. p. 811.
37. Baraza JC, Gracia J, Gil D, Gil PJ. A prototype of a VHDL-based fault injection tool: description and application. J Syst Arch. 2002;47(10):847–67.
38. López C, Entrena L, Olías E. Automatic generation of fault tolerant VHDL designs in RTL. In: FDL (Forum on Design Languages), Lyon, France, September 2001.
39. Carreira J, Madeira H, Silva JG. Xception: software fault injection and monitoring in processor functional units; 1995.
40. Kanawati GA, Kanawati NA, Abraham JA. Ferrari: a flexible software-based fault and error injection system. IEEE Trans Comput. 1995;44(2):248–60.
41. Cotroneo D, Natella R. Fault injection for software certification. In: IEEE security and privacy, special issue on safety-critical systems: the next generation, vol. 11(4). IEEE Computer Society. p. 38–45.
42. Han S, Rosenberg HA, Shin KG. Doctor: an integrated software fault injection environment; 1995.
43. Ziade H, Ayoubi RA, Velazco R. A survey on fault injection techniques. Int Arab J Inf Technol. 2004;1(2):171–86.
44. Miele A. A fault-injection methodology for the system-level dependability analysis of multiprocessor embedded systems. Microprocess Microsyst Embed Hardw Des. 2014;38(6):567–80.
45. de Moraes RLO, Martins E. JACA—a software fault injection tool. In: DSN. IEEE Computer Society; 2003. p. 667.
46. Marinescu PD, Candea G. LFI: a practical and general library-level fault injector. In: DSN. IEEE; 2009. p. 379–88.
47. Dinn AE. Flexible, dynamic injection of structured advice using byteman. In: Proceedings of the tenth international conference on aspect-oriented software development companion, AOSD' 11. New York, NY, USA: ACM; 2011. p. 41–50.
48. Lattner C, Adve V. LLVM: a compilation framework for lifelong program analysis and transformation. In: Proceedings of the international symposium on code generation and optimization: feedback-directed and runtime optimization, CGO '04. Washington, DC, USA: IEEE Computer Society; 2004. p. 75.
49. Kooli M, Benoit P, Di Natale G, Torres L, Sieh V. Fault injection tools based on virtual machines. In: 2014 9th international symposium on reconfigurable and communication-centric systems-on-chip (ReCoSoC), May 2014. p. 1–6.
50. Bellard F. QEMU, a fast and portable dynamic translator. In: Proceedings of the annual conference on USENIX annual technical conference, ATEC '05. Berkeley, CA, USA: USENIX Association; 2005. p. 41.
51. Wan H, Li Y, Xu P. A fault injection system based on QEMU simulator and designed for bit software testing. Appl Mech Mater. 2013;347–350:580–7.
52. Watson J. Virtualbox: bits and bytes masquerading as machines. Linux J. 2008(166).
53. Potyra S, Sieh V, Cin MD. Evaluating fault-tolerant system designs using faumachine. In: Guelfi N, Muccini H, Pelliccione P, Romanovsky A, editors. EFTS. ACM; 2007. p. 9.
54. Breveglieri L, Gueron S, Koren I, Naccache D, Seifert J-P (eds). Fifth international workshop on fault diagnosis and tolerance in cryptography, 2008, FDTC 2008, Washington, DC, USA, 10 August 2008. IEEE Computer Society; 2008.

Chapter 3
Recent Developments in Side-Channel Analysis on Elliptic Curve Cryptography Implementations

Louiza Papachristodoulou, Lejla Batina and Nele Mentens

3.1 Introduction

The emerging need for secure communications in embedded systems is constantly threatened by sophisticated side-channel analysis (SCA) attacks. SCA attacks exploit various types of physical leakage of secret information from cryptographic devices. The physical leakage originates also from the power consumption [1], the electromagnetic radiation [2, 3], and the timing behavior [4] of the device. We focus on attacks exploiting power consumption leakage, namely power analysis attacks. These attacks are based on the principle that a switching event of a signal inside a device causes a current to be drawn from the power supply or to be drained to the ground, which is illustrated in Fig. 3.1 on the basis of a CMOS inverter. When the input switches from a logical 1 to a logical 0 or vice versa, the output makes the opposite transition, respectively charging or discharging the output capacitor. When the input remains constant, there is no switching current and no switching power consumption. This physical behavior is exploited by power analysis attacks to extract data that are processed internally in the device.

Within this area of power analysis of cryptographic implementations, there are various methods of analysis, such as Simple Power Analysis (SPA), Differential Power Analysis (DPA), and Collision Analysis (CA). SPA uses a single power trace or several traces, i.e., the instantaneous power consumption of a single run of an algorithm over a certain period of time. DPA uses statistical methods to extract

L. Papachristodoulou · L. Batina
Digital Security Group, Radboud University, P.O. Box 9010, 6500 Nijmegen, GL, The Netherlands
e-mail: louiza@cryptologio.org

L. Batina
e-mail: lejla@cs.ru.nl

N. Mentens (✉)
KU Leuven, ESAT/COSIC & IMinds, Kasteelpark Arenberg 10, 3001 Leuven, Belgium
e-mail: nele.mentens@kuleuven.be

N. Sklavos et al. (eds.), *Hardware Security and Trust*,
DOI 10.1007/978-3-319-44318-8_3

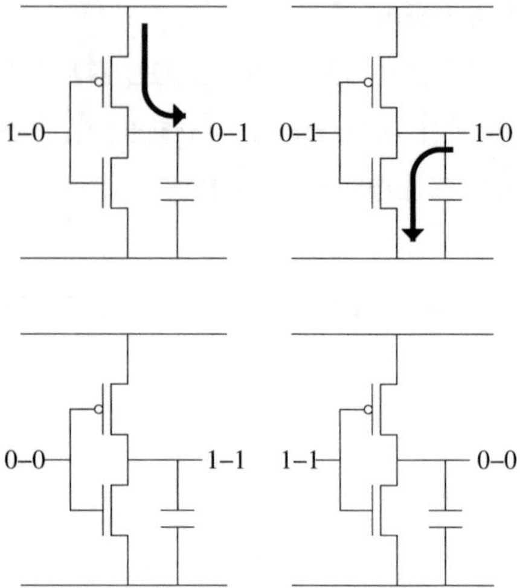

Fig. 3.1 Switching current at the output of a CMOS invertor

information from multiple traces [1]. CA exploits the leakage of two portions of traces when the same intermediate values are used [5].

Naive implementations of public-key cryptosystems are usually susceptible to SPA attacks because of, e.g., the use of conditional branches. In the RSA cryptosystem, these branches are present in the modular exponentiation algorithm when it is executed using an iteration of modular squarings and modular multiplications. The analogy of modular exponentiation in RSA is point multiplication in elliptic curve cryptosystems. Naive implementations use the double-and-add method consisting of consecutive point doublings and point additions, where a point addition is only executed when the corresponding key bit equals 1. This way, a single power trace reveals a logical 1 in the key through the presence of a point addition. One type of countermeasures balance the computation such that the power traces always look similar regardless of the processed key bits. Other countermeasures randomize the computation such that an attacker is not able to correlate the power traces with the processed data.

This chapter starts with an overview of elliptic curves used in cryptography in Sect. 3.2. Since the power analysis attacks we discuss, focus on the scalar multiplication algorithm, Sect. 3.3 presents different options for this algorithm. Section 3.4 elaborates on power analysis attacks on elliptic curve cryptosystems, while Sect. 3.5 gives an overview of countermeasures at the algorithmic level.

3.2 Elliptic Curve Cryptography

Elliptic curve cryptography (ECC) was introduced around 1985 independently by Miller [6] and Koblitz [7]. It is broadly used for implementing asymmetric cryptographic protocols in embedded devices due to the small key length and memory requirements compared to equivalent RSA implementations. For instance, a 256-bit field curve provides a security level of 128 bits, which is roughly equivalent to a 2048-bit RSA key (see [8] for more details).

3.2.1 Coordinate Systems

An elliptic curve $\mathcal{E}$ over the finite field K, denoted as $\mathcal{E}_K$, can be defined in terms of solutions (x, y) to one of the equations defined in Sect. 3.2.2. The pairs that verify these equations represent the *affine coordinates* of a point over the curve $\mathcal{E}$. From the addition rules on an elliptic curve, the necessary operations are addition, multiplication, and inversion over K. Inversion is the most expensive operation and can be avoided using other types of coordinate systems for the points $\mathcal{P} = (x, y)$. We hereby present the most commonly used coordinates systems that can be found in cryptographic implementations of elliptic curve protocols.

Projective coordinates
In the projective coordinate system, each point $\mathcal{P} = (x, y)$ is represented by three coordinates (X, Y, Z), where $x = \frac{X}{Z}, y = \frac{Y}{Z}$.

Jacobian coordinates
In the Jacobian coordinates system, each point $\mathcal{P} = (x, y)$ is represented also by three coordinates (X, Y, Z), with $x = \frac{X}{Z^2}, y = \frac{Y}{Z^3}$.

López-Dahab coordinates
In the López-Dahab system, the relation for the point (X, Y, Z) is $x = \frac{X}{Z}, y = \frac{Y}{Z^2}$.

3.2.2 Forms of Elliptic Curves

There are several forms of elliptic curves defined by their curve equation. Below we will treat some commonly used forms.

Weierstrass curves
An elliptic curve defined over a field K is defined by the Weierstrass equation

$$\mathcal{E}_K : y^2 + \alpha_1 xy + \alpha_3 y = x^3 + \alpha_2 x^2 + \alpha_4 x + \alpha_6. \tag{3.1}$$

Together with the point at infinity $\mathcal{O}$, the set $(\mathcal{E}_K \cup \mathcal{O}, +)$ forms an abelian group with neutral element $\mathcal{O}$.

When the characteristic of the field K is not 2 or 3, then the general Weierstrass form can be simplified to

$$\mathcal{E}_K : y^2 = x^3 + \alpha x + \beta. \tag{3.2}$$

In the following, it is assumed that $char(K) \neq 2, 3$. Adding the points $P = (x_1, y_1)$ and $Q = (x_2, y_2)$ gives a third point on the curve, namely $P + Q = (x_3, y_3)$ according to the formulas

$$\begin{cases} x_3 = \lambda^2 - x_1 - x_2, \\ y_3 = \lambda(x_1 - x_3) - y_1, \end{cases}$$

with $\lambda = \dfrac{y_1 - y_2}{x_1 - x_2}$ if $P \neq Q$ and $\lambda = \dfrac{3x_1^2 + \alpha}{2y_1}$ if $P = Q$.

For points represented in Jacobian coordinates, $P = (X_1, Y_1, Z_1)$ and $Q = (X_2, Y_2, Z_2)$, the addition of P and Q with $P \neq Q$ is $P + Q = (X_3, Y_3, Z_3)$ with

$$X_3 = F^2 - E^3 - 2BE^2, Y_3 = F(BE^2 - X_3) - DE^3, Z_3 = Z_1Z_2E, \tag{3.3}$$

where $A = X_1Z_2^2$, $B = X_2Z_1^2$, $C = Y_1Z_2^3$, $D = Y_2Z_1^3$, $E = A - B$, $F = C - D$. Jacobian addition needs $12\mathbf{M} + 4\mathbf{S}$, with $\mathbf{M}$ and $\mathbf{S}$ the number of multiplications and squarings over K, respectively. Point doubling can be performed very efficiently with only $3\mathbf{M} + 6\mathbf{S}$ using the formulas

$$X_3 = B^2 - 2A, Y_3 = B(A - X_3) - Y_1^4, Z_3 = Y_1Z_1, \tag{3.4}$$

where $A = X_1Y_1^2$, $B = \frac{1}{2}(3X_1^2 + \alpha Z_1^4)$.

Weierstrass curves are standardized and widely used in cryptography [9–12]. However, they have a main drawback regarding their side-channel resistance; namely their addition formulas are incomplete. As is obvious from the previous formulas, addition and doubling are handled differently and the point at infinity gives an exception case. In [13], Bosma and Lenstra presented complete formulas for Weierstrass curves, which had an exceptional case for the pair of points (P, Q) if and only if $P - Q$ is a point of order two. In [14], Renes, Costello, and Batina presented complete addition formulas for odd order elliptic curves $E/\mathbb{F}_q : y^2 = x^3 + \alpha x + b$ with $q \geq 5$, which require only 12 field multiplications $12\mathbf{M}$.[1] The complete addition formulas using Jacobian representation of a point are

$$\begin{cases} X_3 = (X_1Y_2 + X_2Y_1)(Y_1Y_2 - \alpha(X_1Z_2 + X_2Z_1) - 3bZ_1Z_2) \\ \qquad -(Y_1Z_2 + Y_2Z_1)(\alpha X_1X_2 + 3b(X_1Z_2 + X_2Z_1) - \alpha^2 Z_1Z_2), \\ \\ Y_3 = (Y_1Y_2 + \alpha(X_1Z_2 + X_2Z_1) + 3bZ_1Z_2)(Y_1Y_2 - \alpha(X_1Z_2 + X_2Z_1) - 3bZ_1Z_2) \\ \qquad +(3X_1X_2 + \alpha Z_1Z_2)(\alpha X_1X_2 + 3b(X_1Z_2 + X_2Z_1) - \alpha^2 Z_1Z_2), \\ \\ Z_3 = (Y_1Z_2 + Y_2Z_1)(Y_1Y_2 + \alpha(X_1Z_2 + X_2Z_1) + 3bZ_1Z_2) \\ \qquad +(X_1Y_2 + X_2Y_1)(3X_1X_2 + \alpha Z_1Z_2) \end{cases}$$

Edwards curves

Edwards curves, introduced by Edwards in [15], were the first curves shown to have a complete addition law. Applications of Edwards and twisted Edwards curves in

[1] For the overview on elliptic curves in this section, we omit counting the multiplications by a constant (M_α) and the additions $\mathbf{A}$, which are used for extensive comparison results in some publications.

cryptography are extensively studied by Bernstein and Lange [16, 17]. An Edwards curve is defined over a field K with $char(K) \neq 2$ by the following equation:

$$\mathcal{E}_d : y^2 + x^2 = 1 + dx^2y^2, \tag{3.5}$$

where $d \in K\backslash\{0, 1\}$. The Edwards addition law for two points $P = (x_1, y_1)$ and $Q = (x_2, y_2)$, in affine coordinates, is given by the following formulas:

$$P + Q = (x_3, y_3) = (\frac{x_1y_2 + y_1x_2}{1 + dx_1x_2y_1y_2}, \frac{y_1y_2 - x_1x_2}{1 - dx_1x_2y_1y_2}). \tag{3.6}$$

This addition law is unified, i.e., the same formula can be used for both addition and doubling without exceptional cases. The neutral element is the point (0, 1). If d is not a square, then the addition law is complete and there are no exceptional cases for the neutral element.

Twisted Edwards curves, introduced in [18], are a generalization of Edwards curves and they have the form $E_{\alpha,d} : \alpha x^2 + y^2 = 1 + dx^2y^2$. The addition law for twisted Edwards curves is a generalization of Eq. (3.6):

$$P + Q = (x_3, y_3) = (\frac{x_1y_2 + y_1x_2}{1 + dx_1x_2y_1y_2}, \frac{y_1y_2 - \alpha x_1x_2}{1 - dx_1x_2y_1y_2}). \tag{3.7}$$

The cost of addition and doubling on Edwards curves depends on the form of the curve and the coordinates chosen by the developer. An overview of all types of curves and coordinates is given in the Explicit Formulas Database [19]. The most efficient implementation of twisted Edwards curves is given by Bernstein et al. in [18]. It uses inverted twisted Edwards coordinates and needs $9\mathbf{M} + 1\mathbf{S}$ for addition and $3\mathbf{M} + 4\mathbf{S}$ for doubling.[2]

Montgomery curves

In [20], P. L. Montgomery defined the following form of elliptic curves over finite fields of odd characteristic:

$$\mathcal{E}_M : By^2 = x^3 + Ax^2 + x, \quad B(A^2 - 4) \neq 0. \tag{3.8}$$

Let $P_1 = (x_1, y_1)$ and $P_2 = (x_2, y_2)$ be points on $\mathcal{E}_M$. Then, the point $P_3 = (x_3, y_3) = P_1 + P_2$ can be calculated using the following formulas:

Addition formulas ($P_1 \neq P_2$)

$$\begin{cases} \lambda &= (y_2 - y_1)/(x_2 - x_1) \\ x_3 &= B\lambda^2 - A - x_1 - x_2 \\ y_3 &= \lambda(x_1 - x_3) - y_1 \end{cases}$$

[2]The developer can choose to use different formulas for addition and doubling in twisted Edwards curves for extra efficiency in the implementation or unified formulas for resistance against side-channel attacks.

Doubling formulas $(P_1 = P_2)$

$$\begin{cases} \lambda &= (3x_1^2 + 2Ax_1 + 1)/(2By_1) \\ x_3 = B\lambda^2 - A - 2x_1 \\ y_3 = \lambda(x_1 - x_3) - y_1 \end{cases}$$

Montgomery arithmetic is very efficient with additional speed-up by computing only (X, Z) coordinates of intermediate points [21]. We set $(x, y) = (X/Z, Y/Z)$ and present the operations in projective coordinates, as described in [20]. We note here that the point $nP = (X_n, Y_n, Z_n)$ is the n−times multiple of the point $P = (X, Y, Z)$. The addition and doubling formulas for $(m + n)P = mP + nP$ are as follows: **Addition formulas** $(m \neq n)$

$$\begin{cases} X_{m+n} = Z_{m-n}[(X_m - Z_m)(X_n + Z_n) + (X_m + Z_m)(X_n - Z_n)]^2 \\ Z_{m+n} = X_{m-n}[(X_m - Z_m)(X_n + Z_n) - (X_m + Z_m)(X_n - Z_n)]^2 \end{cases}$$

Doubling formulas $(m = n)$

$$\begin{cases} 4X_nZ_n = (X_n + Z_n)^2 - (X_n - Z_n)^2 \\ X_{2n} \quad = (X_n + Z_n)^2(X_n - Z_n)^2 \\ Y_{2n} \quad = (4X_nZ_n)((X_n - Z_n)^2 + ((A + 2)/4)(4X_nZ_n)) \end{cases}$$

In [22], it is shown that the loop iteration in the Montgomery ladder using (X, Z) coordinates is performed in only $11\mathbf{M} + 4\mathbf{S}$. Moreover, as presented in [23], the number of additions and doublings in scalar multiplications on Montgomery form elliptic curves only depends on the bit length of the key and not on the bit patterns or the bit itself at a certain position of the scalar.

Hessian curves

A Hessian curve over a field K is defined by the cubic equation

$$\mathcal{E}_K : x^3 + y^3 + z^3 = dxyz, \tag{3.9}$$

where $d \in K$ and $d^3 \neq 27$. Hessian and twisted Hessian curves are interesting for cryptography due to their small cofactor 3 and their side-channel resistance [24–26]. Moreover, the Hessian addition formulas (also called Sylvester formulas) can be used for doubling, a fact that provides a form of unification.

In [27], Farashahi and Joye presented efficient unified formulas for *generalized* Hessian curves as follows:

$$\mathcal{E}_K : x^3 + y^3 + cz^3 = dxyz, \tag{3.10}$$

where $c, d \in K, c \neq 0$ and $d^3 \neq 27c$. The unified formulas are complete for certain parameter choices. More precisely, the group of K-rational points on a generalized Hessian curve has complete addition formulas, if and only if c is not a cube in K. The fastest known addition formulas on binary elliptic curves with $9\mathbf{M} + 3\mathbf{S}$ for extended projective coordinates and $8\mathbf{M} + 3\mathbf{S}$ for mixed affine-projective addition are presented in [27]. The sum of two points P, Q represented in extended projective coordinates by $(X_i : Y_i : Z_i : A_i : B_i : C_i : D_i : E_i : F_i)$, where $A_i = X_i^2$, $B_i = Y_i^2$, $C_i = Z_i^2$, $D_i = X_iY_i$, $E_i = X_iZ_i$, $F_i = Y_iZ_i$ and for $i = 1, 2$, is the point $P + Q = (X_3 : Y_3 : Z_3 : A_3 : B_3 : C_3 : D_3 : E_3 : F_3)$ with

$$\begin{cases} X_3 = cC_1F_2 + D_1A_2, \\ Y_3 = B_1D_2 + cE_1C_2, \\ Z_3 = A_1E_2 + F_1B_2, \end{cases}$$

with $A_3 = X_3^2$, $B_3 = Y_3^2$, $C_3 = Z_3^2$, $D_3 = X_3Y_3$, $E_3 = X_3Z_3$, $F_3 = Y_3Z_3$.

The complete addition formulas for twisted Hessian curves of cofactor 3 in [24] give the fastest results for prime-field curves with $8.77\mathbf{M}$ for certain curve parameters.

3.3 Scalar Multiplication Algorithms

ECC primitives are used for cryptographic protocols such as the Elliptic Curve Digital Signature Algorithm (ECDSA) for digital signatures, Elliptic Curve ElGamal as an encryption/decryption scheme, and Elliptic Curve Diffie-Hellman (ECDH) as a key exchange scheme. The main operation in all those protocols using ECC is scalar multiplication of a point $\mathcal{P}$ on a curve and with an integer k.

Computing the result of a scalar multiplication on an elliptic curve can be done in a similar way as exponentiation in RSA. A simple and efficient algorithm is binary scalar multiplication, where an n-bit scalar k is written in its binary form $(k_0, k_1, \ldots, k_{n-1})_2$ with $k = \sum_i k_i 2^i$, $k_i \in \{0, 1\}$. The binary algorithm processes a loop, scanning the bits of the scalar (from the most significant bit to the least significant one, or the other way around) and performing a point doubling only if the current bit is 0 or a doubling and an addition if the bit is 1.

Scalar multiplication algorithms take as input a point $\mathcal{P}$ in affine or projective coordinates and the scalar k. The result is the point $[k]\mathcal{P}$ on the curve. During the execution of the algorithm, mixed coordinates can be used for an additional speed-up [28].

Scalar multiplication is a sensitive operation, since it manipulates the secret key k and returns the result according to the bits of the key. Naïve implementations of scalar multiplication with `if`-statements are subject to SCA, and more precisely timing attacks. Coron's randomization countermeasures of point or scalar, as presented in [29] and in Sect. 3.5, can thwart timing or DPA attacks, but not SPA attacks. SPA leakage is present when there is a difference in operation flow between only doubling or doubling-and-addition. Point additions and point doublings give different leakage patterns and since these operations are key dependent, the key can be retrieved quite easily. SPA-resistant algorithms are *regular* algorithms, which perform a constant operation flow regardless of the scalar value. A nice overview of fast and regular scalar multiplication algorithms is given in [30]. In this section, we present the most broadly used regular scalar multiplication algorithms.

3.3.1 Left-to-Right Double-and-Add-Always Algorithm

The double-and-add-always algorithm was initially proposed by Coron in [31] as a first attempt to avoid `if`-statements and therefore prevent the identification of different operations. The algorithm performs a point doubling followed by a point addition in a `for` loop, scanning the scalar bits from the most significant to the least significant one. Both operations are performed in every loop and according to the key bit, the final assignment to $\mathcal{R}_0$ will be either $\mathcal{R}_0$ or $\mathcal{R}_1$. There are no conditional

Algorithm 1: The left-to-right double-and-add-always algorithm

Input: $\mathcal{P}, k = (k_{x-1}, k_{x-2}, \ldots, k_0)_2$
Output: $\mathcal{Q} = k \cdot \mathcal{P}$

$\mathcal{R}_0 \leftarrow \mathcal{P}$;
for $i \leftarrow x-2$ ***down to 0*** **do**
  $\mathcal{R}_0 \leftarrow 2\mathcal{R}_0$;
  $\mathcal{R}_1 \leftarrow \mathcal{R}_0 + \mathcal{P}$;
  $\mathcal{R}_0 \leftarrow \mathcal{R}_{k_i}$;
end

return $\mathcal{R}_0$

statements in the algorithm, but there is one key-dependent assignment, which can leak secret information. Another important remark is that $\mathcal{R}_0$ is initialized by $\mathcal{P}$ instead of $\mathcal{O}$, in order to avoid exceptional cases given by the point at infinity.

3.3.2 Right-to-Left Double-and-Add-Always Algorithm

The binary right-to-left double-and-add-always algorithm of [32] is shown below as Algorithm 2. The steps of the algorithm are similar to Algorithm 1 with the following differences:

- The bits of the scalar are scanned from the least significant to the most significant one.
- Two temporary registers are used instead of three and they are both effectively used, without any dummy operations.

Similar to Algorithm 3.3.1, there are no conditional statements in this algorithm, but there is a key-dependent assignment, which can be vulnerable to attacks. However, there are several attacks that can be mounted on the left-to-right, but not on the right-to-left algorithm (for instance the Doubling attack, described in Sect. 3.4, is only applicable on the left-to-right algorithm).

Algorithm 2: Binary right-to-left double-and-add-always algorithm

Input: $\mathcal{P}, k = (k_{x-1}, k_{x-2}, \ldots, k_0)_2$
Output: $\mathcal{Q} = k \cdot \mathcal{P}$

$\mathcal{R}_0 \leftarrow \mathcal{O}$;
$\mathcal{R}_1 \leftarrow \mathcal{P}$;
for $i \leftarrow 0$ ***up to*** *x-1* **do**
  $b \leftarrow 1 - k_i$;
  $\mathcal{R}_b \leftarrow 2\mathcal{R}_b$;
  $\mathcal{R}_b \leftarrow \mathcal{R}_b + \mathcal{R}_{k_i}$;
end
return $\mathcal{R}_0$

3.3.3 Montgomery Ladder

The Montgomery Ladder, initially presented by Montgomery in [20] as a way to speed up scalar multiplication on elliptic curves, and later used as the primary secure and efficient choice for resource-constrained devices, is one of the most challenging algorithms for simple side-channel analysis due to its natural regularity of operations. A comprehensive security analysis of the Montgomery ladder, given by Joye and Yen in [33], showed that the regularity of the algorithm makes it intrinsically protected against a large variety of implementation attacks (SPA, some fault attacks, etc.). The Montgomery ladder is described in Algorithm 3. For a specific choice of projective coordinates, as described in Sect. 3.2.2 and in [21], one can do computations with only X and Z coordinates, which makes this option more memory efficient than other algorithms.

Algorithm 3: The Montgomery Ladder

Input: $\mathcal{P}, k = (k_{x-1}, k_{x-2}, \ldots, k_0)_2$
Output: $\mathcal{Q} = k \cdot \mathcal{P}$

$\mathcal{R}_0 \leftarrow \mathcal{P}$;
$\mathcal{R}_1 \leftarrow 2\mathcal{P}$;
for $i \leftarrow x - 2$ ***down to*** *0* **do**
  $b \leftarrow 1 - k_i$;
  $\mathcal{R}_b \leftarrow \mathcal{R}_0 + \mathcal{R}_1$;
  $\mathcal{R}_{k_i} \leftarrow 2 \cdot \mathcal{R}_{k_i}$;
end
return $\mathcal{R}_0$

3.3.4 Side-Channel Atomicity

Side-channel atomicity is an SPA countermeasure proposed by Chevallier-Mames et al. [34], in which individual operations are implemented in such a way that they have an identical side-channel profile (e.g., for any branch and any key-bit related subroutine). In short, it is suggested in [34] that the point doubling and addition operations are implemented such that the same code is executed for both operations. This renders the operations indistinguishable by simply inspecting a suitable side-channel. One could, therefore, implement a point multiplication as described in Algorithm 4.

Algorithm 4: Side-Channel Atomic double-and-add algorithm

Input: $\mathcal{P}$, $k = (k_{x-1}, k_{x-2}, \ldots, k_0)_2$
Output: $\mathcal{Q} = k \cdot \mathcal{P}$

$R_0 \leftarrow \mathcal{O}$; $R_1 \leftarrow \mathcal{P}$; $i \leftarrow x - 1$;
$n \leftarrow 0$;
while $i \geq 0$ **do**
  $\mathcal{R}_0 \leftarrow \mathcal{R}_0 + \mathcal{R}_n$;
  $n \leftarrow n \oplus k_i$;
  $i \leftarrow i - \neg n$;
end
return $\mathcal{R}_0$

There are certain choices of coordinates and curves for which this approach can be deployed by using unified or complete addition formulas for the group operations. The unified and complete formulas of Weierstrass, Edwards and Hessian curves are described in Sect. 3.2.2.

3.4 Side-Channel Attacks on ECC

Attacking implementations of elliptic curve cryptography (ECC) with natural protection against side-channel attacks, e.g., implementations using Edwards curves, is quite challenging. This form of elliptic curves, proposed by Edwards in 2007 [15] and promoted for cryptographic applications by Bernstein and Lange [18], showed some advantages compared to elliptic curves in Weierstrass form. For instance, the fast and complete formulas for addition and doubling put these types of curves forward as more appealing for memory-constrained devices and at the same time resistant to classical simple power analysis (SPA) techniques. Recently, due to the work of Renes et al., complete formulas have been published for curves in Weierstrass form [14].

Although considered a very serious threat against ECC implementations, differential power analysis (DPA), as proposed in [1], cannot be applied directly to

ECC-based algorithms and protocols. Soon after the first DPA paper by Kocher et al., Coron showed how to attack the scalar multiplication operation by DPA techniques [31]. However, the idea does not apply to other ECC protocols where the secret is either not a scalar involved in the scalar multiplication algorithm or a scalar that is used only once, like, e.g., in ECDSA or in ephemeral Diffie-Hellman. The latter is incompatible with the requirement of DPA to collect a large number of power traces of computations on the same secret data.

When attacking ECDSA, two secrets could be of interest for the attacker. She could go for an ephemeral key or a secret scalar that becomes a part of the signature. The idea of attacking the ephemeral key is to get reveal a few key bits (from just one measurement) and then proceed with some sort of theoretical cryptanalysis to recover the remaining bits. This kind of special attacks is often used in combination with lattice techniques similar to [35, 36], in order to derive the whole private key from a few bits of multiple ephemeral keys.

The richness of the mathematical structures behind public-key systems and other algorithm-dependent features, that are special for both RSA and ECC, created opportunities for many unique side-channel attacks exploiting those features. The first work to propose new techniques was the paper of Fouque and Vallette [37]. They introduce a new attack against scalar multiplication (or modular exponentiation) that looks for identical patterns within power traces due to the same intermediate results occurring within the computation. In this way, the so-called "doubling attack" only requires two queries to the device. This work has started a new line of research on new attack techniques that reside somewhere between SPA and DPA, of which the most notable are collision [5, 37–41] and template attacks [36, 42, 43].

Collision-based attacks exploit the fact that when processing the same data, the same computations will result in the same (or very similar) patterns in the power consumption traces. However, although the idea is easily verifiable, the efficiency of most of the so far introduced collision-based attacks is shown only on simulated traces; no practical experiments on real ECC implementations have confirmed those results. To the best of our knowledge, only two practical collision-based attacks on exponentiation algorithms were published, each of which rely on very specific assumptions and deal with very special cases. Hanley et al. exploit collisions between input and output operations within the same trace [44]. On the other hand, Wenger et al. performed a hardware-specific attack on consecutive rounds of a Montgomery ladder implementation [45]. However, both attacks are very restrictive in terms of applicability to various ECC implementations as they imply some special implementation options, such as, e.g., the use of López–Dahab coordinates, where field multiplications use the same key-dependent coordinate as input to two consecutive rounds. A class of attacks similar to collision-based attacks is sometimes also called horizontal attacks and they were first defined for modular exponentiation by Clavier et al. [46]. Their attack is inspired by Walter's work [38] and it requires a unique power trace, in this way rendering classical randomization countermeasures. In another work [47], the authors have introduced a general framework enabling to model both horizontal and classical attacks (the latter are called vertical attacks in this work) in a simple way.

Their follow-up paper [41] introduced horizontal attacks for ECC but the results were obtained from simulations only and no real measurements were used.

We observe that the trend of attacks is shifting more toward this type of collision and horizontal attacks as known randomization-based countermeasures are effective in protecting ECC against SPA and DPA attacks. Therefore, in the remainder of this chapter we focus on collision-based/inspired attacks and we give a detailed example of one such attack i.e. the Online Template Attack (OTA).

3.4.1 Collision-Correlation Attacks

As mentioned above, collision attacks exploit leakages by comparing two portions of the same or different traces exploiting the same power being consumed when values are reused. The Big Mac attack [38] is the first theoretical attack on public-key cryptosystems, in which only a single trace is required to observe key dependencies and collisions during an RSA exponentiation. Witteman et al. performed a similar attack on the RSA modular exponentiation even in the presence of blinded messages [48]. Clavier et al. introduced horizontal correlation analysis, as a type of attack where a single power trace is enough to recover the private key [46]. They also extended the Big Mac attack using different distinguishers, i.e., types of statistical tests.

The *doubling attack*, proposed by Fouque and Vallette [37] and described previously, is a special type of collision attack relevant to ECC. The main assumption of this attack is that an adversary can distinguish collisions of power trace segments (within a single or more power traces) when the device under attack performs twice the same computation, even if the adversary is not able to tell which exact computation is done. Collision of two computations will not reveal the value of the operand. Yen et al. extended this attack to the *Refined Doubling Attack (RDA)* [39], where the adversary is assumed to be able to detect the collision between two modular squarings, i.e., detecting if the squared value is the same or not. Collisions of computations cannot be distinguished; the only knowledge obtained is that $k_i = k_{i-1}$ if a collision is detected. Based on the derived relationship between every two adjacent private key bits (either $k_i = k_{i-1}$ or $k_i \neq k_{i-1}$) and a given bit (e.g., k_0 or k_{m-1}), all other private key bits can be derived uniquely. RDA is a powerful attack technique that works against some scalar multiplication algorithms, which are resistant against the doubling attack (e.g., the Montgomery power ladder).

3.4.2 Horizontal Attacks and Variants

An interesting class of side-channel attacks is the *Horizontal Analysis* attack, where a single trace is used to recover the secret scalar. The main characteristic of the traces that makes horizontal attacks possible lies in the fact that the operation sequences of doubling-adding and doubling-doubling can be distinguished. The attacker applies the classical correlation analysis using different parts of time samples in the same side-channel trace to recover the secret scalar bit-by-bit. This technique can be useful to attack protected implementations, where the secret value or unknown input is blinded. The first horizontal attacks were applied to RSA implementations; extension of those to ECC implementations is straight-forward, since scalar multiplication and exponentiation algorithms have the same operation steps.

The so-called *Big Mac* attack from Walter [38] is the first attack of this kind, where squarings (**S**) are distinguished from multiplications (**M**) and the secret exponent of an RSA exponentiation can be recovered from a single execution curve. The distinction is possible by averaging and comparing the cycles performed in the multiplier of the device during long-integer multiplication, since more cycles are needed for **SM** than for **SS**. This attack can be directly applied to ECC implementations.

The term *horizontal* was first introduced by Clavier et al. in [46], where the authors performed a horizontal correlation analysis to compute the correlation factor on several segments extracted from a single execution curve of a known message RSA encryption. More specifically, their proposed method starts by finding a sequence of elementary calculations $(C_i)_j$ (with $i, j \in \mathbb{Z}$ indicating the sequence of the calculation and the order of the execution respectively) that processes the same mathematical operation (e.g., field multiplication) and depends on the same part of the secret scalar. The outputs O_{i_j} of the calculations $C_i(X_i)$ that depend on the same input value X_i will give high correlation results and in this way, they can be distinguished from outputs of computations with different input values. Horizontal correlation analysis was performed on RSA using the Pearson correlation coefficient in [46] and triangular trace analysis of the exponent in [49].

The most recent attack, proposed by Bauer et al. in [41], is a type of *horizontal collision correlation* attack on ECC, which combines atomicity and randomization techniques. Based on the basic assumption of collision attacks that an adversary is able to distinguish when two field multiplications have at least one common operand, their attack consists of the following steps:

- Identify two elementary calculations C_1, C_2 that are processed N times with inputs from the same distribution. The correlation between the random output values O_1, O_2 must depend on the same secret sub-part s.
- For each of the N processings of C_i get an observation l^i_j, with $j \in \{1, \ldots, N\}$.
- Compute the Pearson correlation coefficient on the two samples of observations $\rho = \rho((l^1_j)_j, (l^2_j)_j)$.
- Deduce information on the secret scalar from ρ using an appropriate distinguisher that shows which observation is more similar to the real secret value.

The horizontal collision correlation attack is shown by simulated traces to be applicable to atomic implementations and to implementations based on unified addition formulas over Edwards curves.

Two recent publications on blinded asymmetric algorithms propose the combination of horizontal and vertical techniques, in an attempt to provide more practical attacks against blinded implementations and avoid the complex signal processing phase. Bauer et al. [50] at Indocrypt 2013, presented an attack on RSA blinded exponentiation based on this approach. They took advantage of the side-channel leakage of the entire long-integer modular multiplication without splitting the trace into parts of single precision multiplications. However, their attack requires a small public exponent (no greater than $2^{16}+1$) and an exponent blinding factor smaller than 32 bits. Their observation that the scalar blinding does not mask a large part of the secret value, led Feix et al. [51] a year later to exploit this vulnerability vertically on a ECC implementation. The most significant part of the blinded scalar can be recovered with a horizontal attack. The least significant part of the scalar is retrieved using vertical analysis (several execution traces) and the information leaked in the previous steps of the attack.

3.4.3 Template Attacks

The most powerful SCA attack from an information theoretic point of view is considered to be a template attack (TA). Template attacks, as introduced in the original paper by Chari et al. in [52], are a combination of statistical modeling and power analysis attacks consisting of two phases, as follows:

- The first phase is the *profiling* or *template-building* phase, where the adversary builds templates to characterize the device by executing a sequence of instructions on fixed data. Focusing on an "interesting pattern" or finding the points of interest is very common in this phase.
- The second phase is the *template-matching phase*, in which the adversary matches or correlates the templates to actual traces of the device. By applying some signal processing and classification algorithms to the templates, it is possible to find the best matching for the traces.

In this type of attacks, the adversary is assumed to have in his possession a device which behaves similar to the device under attack (target device), in order to build template traces. In his device he can simulate the same algorithms and implementations that run in the target device. For the template-matching phase several distinguishers and classification algorithms are proposed; in the next section the most common classifiers are presented.

The practical application of TAs is shown on several cryptographic implementations such as RC4 in [53] and elliptic curves in [54]. Medwed and Oswald demonstrated in [42] a practical template attack on ECDSA. However, their attack required an offline DPA attack on the EC scalar multiplication operation during the template-building phase, in order to select the points of interest. They also need 33 template traces per key bit. Furthermore, attacks against ECDSA and other elliptic curve signature algorithms only need to recover a few bits of the ephemeral scalar for multiple scalar multiplications with different ephemeral scalars and can then employ lattice techniques to recover the long-term secret key [35, 36, 43]. It is not possible to obtain several traces in the context of ephemeral Diffie-Hellman: an attacker only gets a single trace and needs to recover sufficiently many bits of this ephemeral scalar from side-channel information to be able to compute the remaining bits through, for example, Kangaroo techniques [55, 56]. With these techniques and by using pre-computation tables, it is possible to exploit partial information on the subkeys and recover the last l unknown bits of the key in $\mathcal{O}(\sqrt[3]{l})$ group operations [57].

3.4.4 Common Distinguishers

In this section, the most common distinguishers used in SCA for correlation analysis and template-matching are presented. Machine learning techniques for classification and clustering are broadly used in SCA, in order to distinguish between traces with high noise ratios.

According to [58], unsupervised clustering is generally useful in side-channel analysis when profiling information is not available and an exhaustive partitioning is computationally infeasible. The authors presented an attack on an FPGA-based elliptic curve scalar multiplication using the $k-$means method. In [59], Perin et al. used unsupervised learning to attack randomized exponentiations.

Lerman et al. showed in [60] that machine learning techniques give better classification results when there is limited ability of the adversary to perform profiling of the device and in a high dimensionality context, where many parameters affect the leakage of the device. Indeed, combining three side-channel leakages and a clustering-based approach for non-profiled attacks, gives higher success rates than traditional template attacks, as shown by Specht et al. in [61].

The success results of online template attacks (presented in the next section) are significantly improved in [62] by using the k-nearest neighbor approach, naïve Bayes classification and the support vector machine method for template classification. In order to explain these techniques, we first give the definition of the Euclidean distance and the Pearson correlation coefficient.

Euclidean Distance The Euclidean distance between two points is defined as the square root of the sum of the squares of the differences between the corresponding point values:

$$d_{EUC} = \sqrt{\sum_{i=1}^{n}(x_i - y_i)^2}$$

In the SCA setting, a realization of a random variable X corresponds to x. A sample of n observations or traces of X is denoted by $(x_i)_{1 \le i \le n}$, where the index i denotes the different observations or the different time when an observation occurs in the same trace.

Pearson correlation The Pearson correlation coefficient measures the linear independence between two observations X and Y:

$$\rho(X,Y)=\frac{n\sum_i x_i y_i-\sum_i x_i\sum_i y_i}{\sqrt{n\sum_i x_i^2-(\sum_i x_i)^2}}\sqrt{n\sum_i y_i^2-(\sum_i y_i)^2}$$

In SCA, the Pearson correlation coefficient is used to describe the difference in the Hamming weight of the observations.

Naïve Bayes Classification The naïve Bayes classification method is based on probability concepts, and more precisely on the Bayes theorem for conditional probabilities of independent events [63]. According to the conditional probability model, let $\mathbf{x}=(x_1,\ldots,x_n)$ be the vector of problem instances (independent variables) to be classified, each one having a feature n. Each instance is assigned a probability $p(c_k|x_1,\ldots,x_n)$, for k possible classes. The set of classes $c_1, c_2, \ldots, c_k$ is mutually exclusive and exhaustive.

Using Bayes' theorem, the posterior conditional probability is $p(c_k|\mathbf{x})=\frac{p(c_k)\ p(\mathbf{x}|c_k)}{p(\mathbf{x})}$. Assuming that each event and posterior probability is independent on each previous event, the conditional distribution over the class variable c is $p(c_k|x_1,\ldots,x_n)=\frac{1}{Z}p(c_k)\prod_{i=1}^{n}p(x_i|c_k)$ where the evidence $Z=p(\mathbf{x})$ is a scaling factor dependent only on $x_1,\ldots,x_n$, that is, a constant if the values of the feature variables are known.

The naïve Bayes classifier is a function that combines the naïve Bayes probability model with a decision rule. One common rule is to pick the hypothesis that is most probable; that is the maximum value of the a posteriori probability. For each class c_i, the class index that gives the maximum value for an event is chosen as classifier.

K-Nearest Neighbor The k-Nearest Neighbor Classification (*kNN*) is a classification method based on the closest instances of the training set to the unlabeled data. Basically, according to [63], it consists of the following two steps:

1. Choose the number of k closest instances (from the training set) to the sample.
2. The majority label (class) for the chosen closest instances will be class for the unlabeled data.

The distance metric plays an important role, in order to determine the closest instance. In kNN, the Euclidean distance or the Manhattan distance can be used. The value k indicates the number of the already-classified closest instances that are chosen in order to classify the next unlabeled data. The default value is 1, but with larger value for k it is possible to obtain higher success rate with less template traces. Figure 3.2 shows an example with $k=2$ and $k=4$ close instances; if the new sample is closer to A, it will be classified in the "A-class."

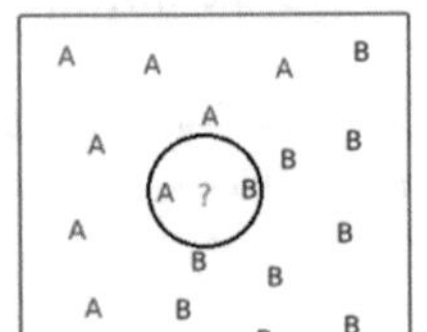

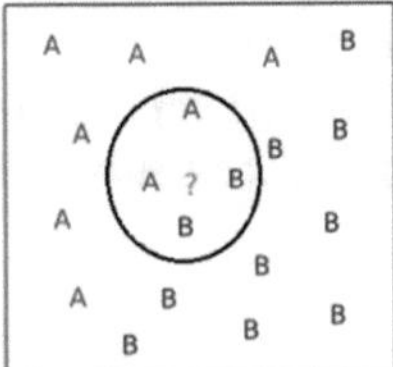

Fig. 3.2 kNN method for $k=2$ and $k=4$

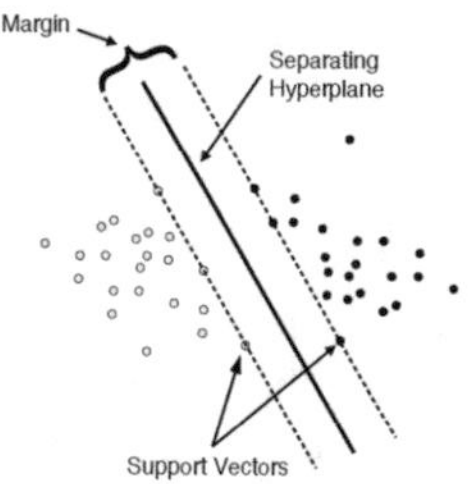

Fig. 3.3 SVM: distance to the hyperplane for two sets of training data

SVM A Support Vector Machine (SVM) is a supervised learning model that produces a discriminative classifier formally defined by a separating hyperplane. In other words, given labeled training data, the algorithm outputs an optimal hyperplane which categorizes new examples. Figure 3.3 shows such a hyperplane. An optimal hyperplane, as defined in [64], is the one that gives the largest minimum distance to the training points, because in this way noisy data will still be classified correctly. Therefore, the optimal separating hyperplane maximizes the margin of the training data. An SVM model is a representation of the examples as points in space, mapped so that the examples of the separate categories are divided by a clear gap that is as wide as possible. New examples are then mapped into that same space and predicted to belong to a category based on which side of the gap they fall on. The classifier in an SVM can be nonlinear for data sets that are not easily separable, but in the following analysis a linear classifier gives very good results.

3.4.5 A Special Case: Online Template Attacks

In this section, we present in more detail an adaptive template attack technique, which is called an *Online Template Attack* (OTA) and is initially proposed by Batina et al. in [65]. This technique resides between horizontal and template attacks. The attacker is able to recover a complete scalar after obtaining only one power trace of a scalar multiplication from the device under attack. This attack is characterized as *online*, because templates are created *after* the acquisition of the target trace. While the same terminology is used, OTA is not a typical template attack; i.e., no preprocessing template-building phase is necessary. OTA functions by acquiring one target trace from the device under attack and comparing patterns of certain operations from this trace with templates obtained from the attacker's device that runs the same implementation. Pattern matching is performed at suitable points in the algorithm, where key-bit related assignments take place by using an automated module based on the Pearson correlation coefficient.

The attacker needs only very limited control over the device used to generate the online template traces. The main assumption is that the attacker can choose the input point to a scalar multiplication, an assumption that trivially holds even without any modification to the template device in the context of ephemeral Diffie-Hellman. It also holds in the context of ECDSA, if the attacker can modify the implementation on the template device or can modify internal values of the computation. This is no different than for previous template attacks against ECDSA.

OTA is defined as a side-channel attack with the following conditions:

1. The attacker obtains only one power trace of the cryptographic algorithm involving the targeted secret data. This trace is called the *target trace*. The device, from which the target trace is obtained, is the *target device*. The fact that only one target trace is necessary for the attack, makes it possible to attack scalar multiplication algorithms with ephemeral scalar and with randomized scalar.
2. The attacker is generating template traces *after* having obtained the target trace. These traces are called *(online) template traces*.
3. The attacker obtains the template traces on the target device or a similar device[3] *with very limited control over it*, i.e., access to the device to run several executions with chosen public inputs. The attacker does not rely on the assumption that the secret data are the same for all template traces.
4. At least one assignment in the exponentiation algorithm is made depending on the value of particular scalar bit(s), but there are no branches with key-dependent computations. Since the doubling operation is attacked, this key-dependent assignment should be during doubling. As a counterexample, it is noted that the binary right-to-left add-always algorithm for Lucas recurrences [32] is resistant to the proposed attack, because the result of the doubling is stored in a non-key-dependent variable.

The attack methodology is as follows:

- Acquire a full target trace from the device under attack, during the execution of a scalar multiplication.
- Locate the doubling and addition operations performed in each round.
- Find multiples of $m\mathcal{P}$, where $m \in \mathbb{Z}$, $m \leq k$ and k is the scalar. These points are used to create the template traces.

The methodology offers a generic attack framework, which does not require any previous knowledge of the leakage model nor a specific type of curve. It is applicable to various forms of curves (Weierstrass, Edwards and Montgomery curves), scalar multiplication algorithms and implementations. Contrary to the doubling attack [37], OTA can be launched against right-to-left algorithms and the Montgomery ladder.

The basic idea, as depicted in Fig. 3.4 consists of comparing the traces for the inputs $\mathcal{P}$ (target trace) and $2\mathcal{P}$ (online template trace) while executing scalar multiplication and then finding similar patterns between them, based on the hypothesis on a bit for a given operation. The target trace is obtained only once. For every bit of the scalar, an online template trace with input $k\mathcal{P}$, $k \in \mathbb{Z}$ is obtained, where k is chosen as a function of a hypothesis on this bit.

In the original paper, pattern matching is performed using the Pearson correlation coefficient, $\rho(X, Y)$, which measures the linear relationship between two vari-

[3] Similar device means the same type of microcontroller running the same algorithm.

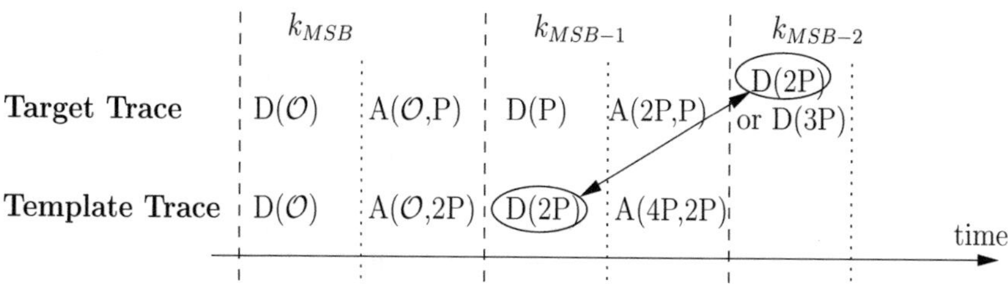

Fig. 3.4 Schematic representation of OTA [66]

ables X and Y. For power traces, the correlation coefficient shows the relationship between two points of the trace, which indicates the Hamming-weight leakage of key-dependent assignments during the execution of a cryptographic algorithm. Extension to other distinguishers from machine learning is performed in [62].

The template matching corresponds to a list of correlation coefficients that show the relationship between all samples from the template trace to the same consecutive amount of samples in the target trace. If the hypothesis on the given key bit is correct, then the pattern match between the template traces at the targeted operation will be high (in experiments it reached 99 %).

In this way, the first i bits of the key can be recovered. Knowledge of the first i bits provides the attacker with complete knowledge of the internal state of the algorithm just before the $(i+1)$th bit is processed. Since at least one operation in the loop depends on this bit, a hypothesis can be made about the $(i+1)$th bit, a template trace based on this hypothesis is computed, and this trace is correlated with the target trace at the relevant predetermined point of the algorithm.

OTA on scalar multiplication algorithms The core idea and feasibility of the attack is demonstrated through an example based on the double-and-add-always algorithm described in Algorithm 1. Table 3.1 shows two executions of the algorithm for two different scalars $k = 100$ and $k = 110$. The first execution of the loop always starts by doubling the input point $\mathcal{P}$, for all values of k. It is assumed that $k_{x-1} = 1$. Depending on the second-most significant key bit k_{x-2}, the output of the first iteration of the algorithm will be either $2\mathcal{P}$ or $3\mathcal{P}$. For any point $\mathcal{P}$ it is, therefore, possible to get a power trace for the operation $2\mathcal{P}$, i.e., the attacker lets the algorithm execute the first two double-and-add iterations. In the proposed setup, the authors could zoom into the level of one doubling, which will be the template trace. Then, the attacker can

Table 3.1 Two executions of the double-and-add-always algorithm [65]

$k = 100$	$k = 110$
$R_0 = P$	$R_0 = P$
$R_0 = 2P$, $R_1 = 3P$, return $2P$	$R_0 = 2P$, $R_1 = 3P$, return $3P$
$R_0 = 4P$, $R_1 = 5P$, return $4P$	$R_0 = 6P$, $R_1 = 7P$, return $6P$

perform the same procedure with $2\mathcal{P}$ as the input point to obtain the online template trace that he wants to compare with the target trace. If it is assumed, that the second-most significant bit of k is 0, then he compares the $2\mathcal{P}$ template with the output of the doubling in the first iteration. Otherwise, he compares it with the online template trace for $3\mathcal{P}$.

Assuming that the first $(i-1)$ bits of k are known, he can derive the ith bit by computing the two possible states of $\mathcal{R}_0$ after this bit has been treated and recover the key iteratively. Note that only the assignment in the ith iteration depends on the key bit k_i, but none of the computations do, so it is necessary to compare the trace of the doubling operation in the $(i+1)$th iteration with the original target trace. To decide whether the ith bit of k is zero or one, the trace that the doubling operation in the $(i+1)$th iteration would give for $k_{i+1}=0$ is compared with the target trace. For completeness, he can compare the target trace with a trace obtained for $k_{i+1}=1$ and verify that it has a lower pattern match percentage; in this case, the performed attack needs two template traces per key bit. However, if during the acquisition phase the noise level is low and the signal is of good quality, an efficient attack can be performed with only the target trace and a single trace for the hypothetical value of $\mathcal{R}_{k_{i+1}}$.

Attacking the right-to-left double-and-add-always algorithm of [32] can be done in a similar way, since it is a type of key-dependent assignment OTA. The attacker targets the doubling operation and notes that the input point will be doubled either in the first (if $k_0=0$) or in the second iteration of the loop (if $k_0=1$). If k is fixed he can easily decide between the two by inputting different points, since if $k_0=1$ he will see the common operation $2\,\mathcal{O}$. If k is not fixed, he simply measures the first two iterations and again uses the operation $2\,\mathcal{O}$ if the template generator should use the first or second iteration. Once he is able to obtain clear traces, the attack itself follows the general description of an OTA.

Montgomery Ladder The main observation that makes OTA attacks applicable to the Montgomery ladder is that at least one of the computations, namely the doubling in the main loop, directly depends on the key bit k_i. For example, if it is assumed that the first three bits of the key are 100, then the output of the first iteration will be $R_0=2\mathcal{P}$. If it is assumed that the first bits are 110, then the output of the first iteration will be $R_0=3\mathcal{P}$. Therefore, if the attacker compares the pattern of the output of the first iteration of Algorithm 3 with scalar $k=100$, he will observe a higher correlation with the pattern of $R_0=2\mathcal{P}$ than with the pattern of $R_0=3\mathcal{P}$. This is demonstrated in the working example of Table 3.2.

Table 3.2 Two executions of the Montgomery ladder [65]

$k=100$	$k=110$
$R_0=P,\ R_1=2P$	$R_0=P,\ R_1=2P$
$b=1\quad R_1=3P,\ R_0=2P$	$b=0\quad R_0=3P,\ R_1=4P$
$b=1\quad R_1=5P,\ R_0=4P$	$b=1\quad R_1=7P,\ R_0=6P$

Side-channel atomicity

Simple atomic algorithms do not offer any protection against online template attacks, because the regularity of point operations does not prevent mounting this sort of attack. The point $2\mathcal{P}$, as an output of the third iteration of Algorithm 4, will produce a power trace with a pattern that is very similar to the trace that would have the point $2\mathcal{P}$ as an input. Therefore, the attack will be the similar to the one described for the binary left-to-right double-and-add-always algorithm; the only difference is that instead of the output of the second iteration of the algorithm, the attacker has to focus on the pattern of the third iteration. In general, when an attacker forms a hypothesis about a certain number of bits of k, the hypothesis will include the point in time where $\mathcal{R}_0$ will contain the predicted value. This means that he would have to acquire a larger target trace to allow all hypotheses to be tested.

Practical results The feasibility and efficiency of OTA is shown in [65] with practical attacks on the double-and-add-always scalar multiplication running on the ATmega163 microcontroller [67] in a smart card. The scalar multiplication algorithm is based on the curve arithmetic of the Ed25519 implementation presented in [68], which is available online at http://cryptojedi.org/crypto/#avrnacl. The elliptic curve used in Ed25519 is the twisted Edwards curve $E : -x^2 + y^2 = 1 + dx^2y^2$ with $d = -(121665/121666)$ and base point

$$\mathcal{P} = (\texttt{15112221349535400772501151409588531511454012693041857206046113283949847762202},$$
$$\texttt{46316835694926478169428394003475163141307993866256225615783033603165251855960}).$$

For more details on Ed25519 and this specific curve, see [69, 70]. The whole underlying field and curve arithmetic is the same as in [68]. This means in particular that points are internally represented in extended coordinates as proposed in [71]. In this coordinate system, a point $\mathcal{P} = (x, y)$ is represented as $(X : Y : Z : T)$ with $x = X/Z$, $y = Y/Z$, and $x \cdot y = T/Z$.

Experimental results of OTA with extended projective coordinates of 256 bits, extended projective coordinates with reduced 255-bit input and input points with affine compressed coordinates are presented in [65]. The attack targets the output of the doubling operation and then performs pattern matching based on the Pearson correlation coefficient.

The correct key bit guess ($k_2 = 0$) gives 97 % correlation of the target trace with the template trace for $2\mathcal{P}$. On the other hand, the correlation of the target trace with the template trace for $3\mathcal{P}$ is at most 83 %. These high correlation results hold when one key bit is attacked. For every key bit, the pattern matching will give peaks as in Fig. 3.5. Attacking five bits with one acquisition gives lower numbers for pattern matching for both the correct and the wrong scalar guess, mainly due to the noise that is higher for longer acquisitions. However, the difference between correct and wrong assumptions is still remarkable; correct bit assumptions have 84–88 % matching patterns, while the correlation for the wrong assumptions drops to 50–72 %. To determine the value of one bit, it is thus necessary to compute only one template trace, and decide on the value of the targeted bit depending on whether the

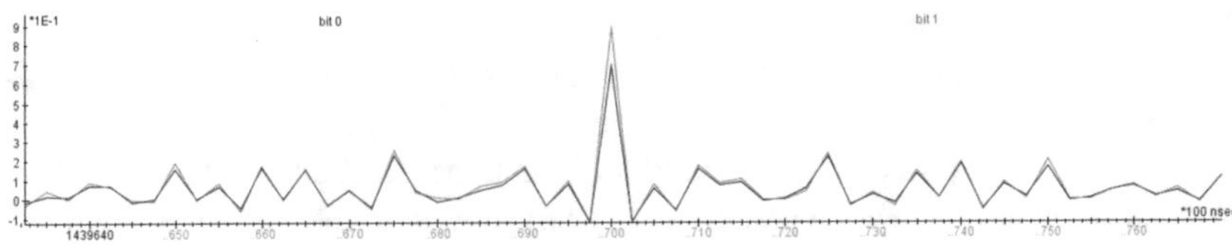

Fig. 3.5 Pattern Matching $2\mathcal{P}$ (*blue*) to target and $3\mathcal{P}$ to target (*brown*) [65]

correlation is above or below a certain threshold (in this case, the threshold can be set to 80 %.

Error-Detection and Correction The idea of Online Template Attacks was extended by Dugardin et al. in [66] with an adaptive template attack on scalar multiplication. The authors propose a generic method to distinguish matching templates and using two templates per key bit, they manage to detect and correct errors for wrong bit assumptions. This fact increases the success rate of the attack significantly compared to the original OTA, reaching 99.8 % when 100 average template traces are used. They also take advantage of the horizontal and vertical leakage, which occurs in the broadly used software implementation of mbedTLS during the modular multiplication of large numbers (256-bit elements).

Classification Algorithms for Template Attacks The fact that the template-building phase in OTA is not necessary, significantly simplifies the process of retrieving the key, leaving the overhead of the attack in the template-matching phase. The template-matching technique used for both OTA papers [65, 66] is based on the Pearson correlation coefficient. In [62], more efficient techniques from the field of Machine Learning are used as distinguishers, and the proposed attack reaches a success rate of 100 % with only 20 template traces per key bit. This work is the first step toward a framework for "automating" the template-matching phase. The attack can be classified as a form of OTA having the same attack model and assumptions. The proposed classification techniques from the field of Machine Learning (k−Nearest Neighbor, Naïve Bayes, SVM) provide an efficient and simplified way to match templates during a Template Attack with very high success rates. A practical application of this attack is demonstrated on the scalar multiplication algorithm for the Brainpool curve BP256r1 implemented in mbedTLS (formerly PolarSSL, version 1.3.7).

3.5 Countermeasures

To prevent first-order DPA attacks [1, 4], it is not sufficient to make the operations time-constant and the power traces indistinguishable. The most common countermeasure applied in ECC implementations is randomization of the secret values. In this way, developers make it more difficult to extract useful information from secret values. This section first covers different types of randomization. Further, we focus specifically on countermeasures against OTA attacks.

3.5.1 Randomization Countermeasures

Scalar randomization Instead of a point multiplication with the scalar k, the blinded scalar k' is used, which is computed as follows:

$$k' = k + \#E \cdot r\,.$$

Here, $\#E$ is the number of points on the curve and r is a random number [31]. Because kP and $k'P$ always result in the same point on the elliptic curve, this method is effective against first-order DPA attacks when the random number is changed for every execution of the point multiplication.

Projective coordinate randomization In addition to scalar randomization, another countermeasure against DPA attacks on elliptic curve point multiplication is projective coordinate randomization. This countermeasure exploits the fact that the Z-coordinate can be chosen randomly when using projective coordinates [31]. This comes down to choosing a different Z-coordinate for each point multiplication during the conversion of the input point P to projective coordinates.

Base point splitting Using this technique, the scalar multiplication is not performed on the point P, but on the point $P + R$, where R is a random point on the curve. After the point multiplication $k(P + R)$, the value kR is subtracted from the result.

Elliptic curve isomorphism randomization The idea to protect scalar multiplication by transforming a curve through various random morphisms, was initially proposed by Joye and Tymen in [72]. Assume that ϕ is a random isomorphism from $\mathcal{E}_K \to \mathcal{E}'_K$, which maps $\mathcal{P} \in \mathcal{E}_K \to \mathcal{P}' \in \mathcal{E}'_K$. Multiplying $\mathcal{P}'$ with k will give $\mathcal{Q}' = [k]\mathcal{P}' \in \mathcal{E}'_K$. With the inverse map ϕ^{-1} we can get back to $\mathcal{Q} = [k]\mathcal{P}$. An attacker needs to know the internal representation of the point in order to perform a successful attack, so if $\mathcal{P}'$ is on a curve that the adversary does not know, he cannot create input points in the correct representation.

3.5.2 OTA Countermeasures

Given that an attacker needs to predict the intermediate state of an algorithm at a given point in time, we can assume that the countermeasures that are used to prevent DPA will also have an effect on the OTA. There are methods for changing the representation of a point, which can prevent OTA and make the result unpredictable to the attacker. Most notably those countermeasures are randomizing the projective representation of points and randomizing the coordinates through a random field isomorphism as described in [73]. However, inserting a point in affine coordinates and changing to (deterministic) projective coordinates during the execution of the scalar multiplication (compressing and decompressing of a point), does not affect the OTA type of attack, as it is shown with practical experiments in [65].

References

1. Kocher P, Jaffe J, Jun B. Differential power analysis. In: Wiener M, editor. Advances in cryptology CRYPTO '99, vol. 1666. LNCS, Springer; 1999. p. 388–97.
2. Gandolfi K, Mourtel C, Olivier F. Electromagnetic analysis: concrete results. In: Proceedings of third international work-shop, Cryptographic hardware and embedded systems—CHES 2001, Paris, France, May 14–16, 2001. Generators; 2001. p. 251–61. doi:10.1007/3-540-44709-1_21. http://dx.doi.org/10.1007/3-540-44709-1_21.
3. Quisquater J-J, Samyde D. Electro magnetic analysis (EMA): measures and counter-measures for smart cards. In: Proceedings of the international conference on research in smart cards: smart card programming and security. E-SMART '01. London, UK, UK: Springer; 2001. p. 200–10. ISBN:3-540-42610-8. http://dl.acm.org/citation.cfm?id=646803.705980.
4. Kocher PC. Timing attacks on implementations of Diffe-Hellman, RSA, DSS, and other systems. In: Koblitz N, editor. Advances in cryptology CRYPTO '96, vol. 1109. LNCS, Springer; 1996. p. 104–13.
5. Schramm K, Wollinger T, Paar C. A new class of collision attacks and its application to DES. In: Johansson T, editor. Fast software encryption, vol. 2887. LNCS, Springer; 2003. p. 206–22.
6. Miller VS. Use of elliptic curves in cryptography. In: Williams HC, editor. Proceedings of advances in cryptology—CRYPTO '85, Santa Barbara, California, USA, August 18–22, 1985, vol. 218. Lecture notes in computer science. Springer; 1985. p. 417–26. ISBN:3-540-16463-4. doi:10.1007/3-540-39799-X_31. http://dx.doi.org/10.1007/3-540-39799-X_31.
7. Koblitz N. Elliptic curve cryptosystems. Math Comput. 1987;48:203–9.
8. Blake NSI, Seroussi G. Advances in elliptic curve cryptography, vol. 317. Cambridge University Press; 1999.
9. ANSI-X9.62. Public key cryptography for the financial services industry: the elliptic curve digital signature algorithm (ECDSA), 1998.
10. ANSI-X9.63. Public key cryptography for the financial services industry: key agreement and key transport using elliptic curve cryptography, 1998.
11. BSI. RFC 5639—Elliptic curve cryptography (ECC) brainpool standard curves and curve generation. Technical report Bundesamt für Sicherheit in der Informationstechnik (BSI), 2010.
12. NIST. FIPS Publication 186-4—Digital signature standard (DSS). Tech. rep. National Institute of Standards and Technology (NIST), 2013. http://nvlpubs.nist.gov/nistpubs/FIPS/NIST.FIPS.186-4.pdf.
13. Bosma W, Lenstra H. Complete systems of two addition laws for elliptic curves. J Number Theory 1995;53(2):229–40. ISSN:0022-314X. http://dx.doi.org/10.1006/jnth.1995.1088. http://www.sciencedirect.com/science/article/pii/S0022314X85710888.
14. Renes J, Costello C, Batina L. Complete addition formulas for prime order elliptic curves. In: Fischlin M, Coron J-S, editors. Proceedings of progress in cryptology EUROCRYPT 2016 (35th international conference on cryptology in Europe, Vienna, Austria, May 8–12, 2016), vol. 9665. LNCS, Springer. p. 403–28.
15. Edwards HM. A normal form for elliptic curves. In: Koç ÇK, Paar C, editors. Bulletin of the American mathematical society, vol. 44. 2007. p. 393–422. http://www.ams.org/journals/bull/2007-44-03/S0273-0979-07-01153-6/home.html.
16. Bernstein DJ, Lange T. A complete set of addition laws for incomplete Edwards curves. In: IACR cryptology ePrint archive 2009. p. 580. http://eprint.iacr.org/2009/580.
17. Bernstein DJ, Lange T. Faster addition and doubling on elliptic curves. In: Kurosawa K, editor. Advances in cryptology ASIACRYPT 2007, vol. 4833. LNCS, Springer; 2007. p. 29–56. http://cr.yp.to/papers.html#newelliptic.
18. Bernstein DJ, Birkner P, Joye M, Lange T, Peters C. Twisted edwards curves. In: Vaudenay S, editor. Progress in cryptology AFRICACRYPT 2008, vol. 5023. LNCS, Springer; 2008, p. 389–405. http://cr.yp.to/papers.html/#twisted.
19. Bernstein DJ, Lange T. Explicit formulas database. http://www.hyperelliptic.org/EFD/.
20. Montgomery PL. Speeding the pollard and elliptic curve methods of factorization. Math Comput. 1987;48(177):243–64.

21. Stam M. On montgomery-like representations for elliptic curves over GF(2k). In: Desmedt YG, editor. Proceedings of public key cryptography PKC 2003: 6th international workshop on practice and theory in public key cryptography Miami, FL, USA, January 6–8, 2003. Berlin, Heidelberg: Springer; 2002. p. 240–54. ISBN: 978-3-540-36288-3. doi:10.1007/3-540-36288-6_18. http://dx.doi.org/10.1007/3-540-36288-6_18.
22. Izu T, Möller B, Takagi T. Improved elliptic curve multiplication methods resistant against side channel attacks. In: Progress in cryptology—INDOCRYPT 2002, third international conference on cryptology in India, Hyderabad, India, December 16–18, 2002. p. 296–313. doi:10.1007/3-540-36231-2_24. http://dx.doi.org/10.1007/3-540-36231-2_24.
23. Okeya K, Kurumatani H, Sakurai K. Elliptic curves with the montgomery-form and their cryptographic applications. In: Proceedings public key cryptography, third international workshop on practice and theory in public key cryptography, PKC 2000, Melbourne, Victoria, Australia, January 18–20, 2000. p. 238–57. doi:10.1007/978-3-540-46588-1_17. http://dx.doi.org/10.1007/978-3-540-46588-1_17.
24. Bernstein DJ, Chuengsatiansup C, Kohel D, Lange T. Twisted hessian curves. In: Proceedings of progress in cryptology—LATINCRYPT 2015—4th international conference on cryptology and information security in Latin America, Guadalajara, Mexico, August 23–26, 2015. p. 269–94. doi:10.1007/978-3-319-22174-8_15. http://dx.doi.org/10.1007/978-3-319-22174-8_15.
25. Hisil H, Wong KK-H, Carter G, Dawson E. Faster group operations on elliptic curves. In: Brankovic L, Susilo W, editors. Seventh Australasian information security conference (AISC 2009), vol. 98. CRPIT. Wellington, New Zealand: ACS; 2009. p. 7–19.
26. Joye M, Quisquater J. Hessian elliptic curves and side-channel attacks. In: Proceedings of Cryptographic hardware and embedded systems—CHES 2001, third international workshop, Paris, France, May 14–16, 2001. Generators, 2001. p. 402–10. doi:10.1007/3-540-44709-1_33. http://dx.doi.org/10.1007/3-540-44709-1_33.
27. Farashahi RR, Joye M. Effcient arithmetic on hessian curves. In: Proceedings of Public key cryptography—PKC 2010, 13th international conference on practice and theory in public key cryptography, Paris, France, May 26–28, 2010. p. 243–60. doi:10.1007/978-3-642-13013-7_15. http://dx.doi.org/10.1007/978-3-642-13013-7_15.
28. Cohen H, Miyaji A, Ono T. Efficient elliptic curve exponentiation using mixed coordinates. In: Ohta K, Pei D, editors. Proceedings of advances in cryptology—ASIACRYPT '98, international conference on the theory and applications of cryptology and information security, Beijing, China, October 18–22, 1998. Lecture notes in computer science, vol. 1514. Springer; 1998. p. 51–65. ISBN: 3-540-65109-8. doi:10.1007/3-540-49649-1_6. http://dx.doi.org/10.1007/3-540-49649-1_6.
29. Coron J. Resistance against differential power analysis for elliptic curve cryptosystems. In: Koç ÇK, Paar C, editors. Proceedings of cryptographic hardware and embedded systems, first international workshop, CHES '99, Worcester, MA, USA, August 12–13, 1999. Lecture notes in computer science, vol. 1717. Springer; 1999. p. 292–302. ISBN: 3-540-66646-X. doi:10.1007/3-540-48059-5_25. http://dx.doi.org/10.1007/3-540-48059-5_025.
30. Rivain M. Fast and regular algorithms for scalar multiplication over elliptic curves. In: IACR Cryptology ePrint Archive, 2011. p. 338. http://eprint.iacr.org/2011/338.
31. Coron J-S. Resistance against differential power analysis for elliptic curve cryptosystems. In: Koç ÇK, Paar C, editors. Cryptographic hardware and embedded systems CHES'99, vol. 1717. LNCS, Springer; 1999. p. 292–302. http://saluc.engr.uconn.edu/refs/sidechannel/coron99resistance.pdf.
32. Joye M. Highly regular right-to-left algorithms for scalar multiplication. In: Paillier P, Verbauwhede I, editors. Cryptographic hardware and embedded systems CHES 2007, vol. 4727. LNCS, Springer; 2007. p. 135–47.
33. Joye M, Yen S. The montgomery powering ladder. In: Kaliski BS, Koç ÇK, Paar C, editors. Cryptographic hardware and embedded systems CHES 2002, vol. 2523. LNCS, Springer; 2002. p. 291–302.
34. Chevallier-Mames B, Ciet M, Joye M. Low-cost solutions for preventing simple sidechannel analysis: side-channel atomicity. IEEE Trans Comput. 2004;53(6):760–8.

35. Benger N, van de Pol J, Smart NP, Yarom Y. Ooh aah... just a little bit: a small amount of side channel can go a long way. In: Proceedings of cryptographic hardware and embedded systems—CHES 2014—16th international workshop, Busan, South Korea, September 23–26, 2014. p. 75–92. doi:10.1007/978-3-662-44709-3_5. http://dx.doi.org/10.1007/978-3-662-44709-3_5.
36. Römer T, Seifert J. Information leakage attacks against smart card implementations of the elliptic curve digital signature algorithm. In: Attali I, Jensen T, editors. Smart card programming and security, vol. 2140. LNCS, Springer; 2001. p. 211–19.
37. Fouque P-A, Valette F. The doubling attack why upwards is better than downwards. In: Walter CD, Koç ÇK, Paar C, editors. Cryptographic hardware and embedded systems CHES 2003, vol. 2779. LNCS, Springer; 2003. p. 269–80.
38. Walter CD. Sliding windows succumbs to big mac attack. In: Koç ÇK, Naccache D, Paar C, editors. Cryptographic hardware and embedded systems CHES 2001, vol. 2162. LNCS, Springer; 2001. p. 286–99.
39. Yen S, Ko L, Moon S, Ha J. Relative doubling attack against montgomery ladder. In: Won DH, Kim S, editors. Information security and cryptology ICISC 2005, vol. 3935. LNCS, Springer; 2005. p. 117–28.
40. Homma N, Miyamoto A, Aoki T, Satoh A, Shamir A. Collision-based power analysis of modular exponentiation using chosen-message pairs. In: Oswald E, Rohatgi P, editors. Cryptographic hardware and embedded systems—CHES 2008, vol. 5154. LNCS, Springer; 2008. p. 15–29.
41. Bauer A, Jaulmes É, Prouff E, Wild J. Horizontal collision correlation attack on elliptic curves. In: Lange T, Lauter K, Lisonek P, editors. Selected areas in cryptography, vol. 8282. LNCS, Springer; 2014. p. 553–70.
42. Medwed M, Oswald E. Template attacks on ECDSA. In: Chung K-I, Sohn K, Yung M, editors. Information security applications, vol. 5379. LNCS, Springer; 2009. p. 14–27.
43. Mulder ED, Hutter M, Marson ME, Pearson P. Using bleichenbacher's solution to the hidden number problem to attack nonce leaks in 384-Bit ECDSA. In: Bertoni G, Coron J-S, editors. Cryptographic hardware and embedded systems CHES 2013, vol. 8086. LNCS, Springer; 2013. p. 435–52. https://online.tugraz.at/tug_online/voe_main2.getvolltext?pCurrPk=71281.
44. Hanley N, Kim H, Tunstall M. Exploiting collisions in addition chain-based expo-nentiation algorithms using a single trace. Cryptology ePrint Archive, Report 2012/485.2012.
45. Wenger E, Korak T, Kirschbaum M. Analyzing side-channel leakage of RFID suitable lightweight ECC hardware. In: Hutter M, Schmidt J-M, editors. Radio frequency identification, vol. 8262. LNCS, Springer; 2013. p. 128–44.
46. Clavier C, Feix B, Gagnerot G, Roussellet M, Verneuil V. Horizontal correlation analysis on exponentiation. In: Soriano M, Qing S, Lopez J, editors. Information and communications security, vol. 6476. LNCS, Springer; 2010. p. 46–61.
47. Bauer A, Jaulmes É, Prouff E, Wild J. Horizontal and vertical side-channel attacks against secure RSA implementations. In: Proceedings topics in cryptology—CT-RSA 2013—the cryptographers' track at the RSA conference 2013, San Francisco,CA, USA, February 25–March 1, 2013. p. 1–17. doi:10.1007/978-3-642-36095-4_1. http://dx.doi.org/10.1007/978-3-642-36095-4_1.
48. Witteman M, van Woudenberg J, Menarini F. Defeating RSA multiply-always and message blinding countermeasures. In: Kiayias A, editor. Topics in cryptology CT-RSA 2011, vol. 6558. LNCS, Springer; 2011. p. 77–88.
49. Clavier C, Feix B, Gagnerot G, Giraud C, Roussellet M, Verneuil V. ROSETTA for single trace analysis. In: Galbraith S, Nandi M, editors. Progress in cryptology INDOCRYPT 2012, vol. 7668. LNCS, Springer; 2012. p. 140–55.
50. Bauer A, Jaulmes É. Correlation analysis against protected SFM implementations of RSA. In: Proceedings progress in cryptology—INDOCRYPT 2013—14th international conference on cryptology in India, Mumbai, India, December 7–10, 2013. p. 98–115. doi:10.1007/978-3-319-03515-4_7. http://dx.doi.org/10.1007/978-3-319-03515-4_7.
51. Feix B, Roussellet M, Venelli A. Side-channel analysis on blinded regular scalar multiplications. Cryptology ePrint Archive, Report 2014/191. http://eprint.iacr.org/.2014.

52. Chari S, Rao JR, Rohatgi P. Template attacks. In: Cryptographic hardware and embedded systems—CHES 2002, 4th international workshop, Redwood Shores, CA, USA, August 13–15, 2002, Revised Papers. 2002. p. 13–28. doi:10.1007/3-540-36400-5_3. http://dx.doi.org/10.1007/3-540-36400-5_3.
53. Rechberger C, Oswald ME. Practical template attacks. In: Lim CH, Yung M, editors. Information security applications, vol. 3325. Lecture notes in computer science. Springer; 2004. p. 440–56.
54. Mulder ED, Buysschaert P, Örs SB, Delmotte P, Preneel B, Vandenbosch G, Verbauwhede I. Electromagnetic analysis attack on an FPGA implementation of an elliptic curve cryptosystem. In: IEEE international conference on computer as a tool. Belgrade, Serbia & Montenegro; 2005. p. 1879–82. doi:10.1109/EURCON.2005.1630348. http://www.sps.ele.tue.nl/members/m.j.bastiaans/spc/demulder.pdf.
55. Avanzi RM. Generic algorithms for computing discrete logarithms. In: Handbook of elliptic and hyperelliptic curve cryptography, 2005. p. 476–94. doi:10.1201/9781420034981.pt5. http://dx.doi.org/10.1201/9781420034981.pt5.
56. Pollard JM. Kangaroos, monopoly and discrete logarithms. J. Crypt. 2000;13(4):437–47. doi:10.1007/s001450010010. http://dx.doi.org/10.1007/s001450010010.
57. Lange T, van Vredendaal C, Wakker M. Kangaroos in side-channel attacks. In:Smart card research and advanced applications—13th international conference, CARDIS 2014, Paris, France, November 5–7, 2014. Revised selected papers. 2014. p. 104–21. doi:10.1007/978-3-319-16763-3_7. http://dx.doi.org/10.1007/978-3-319-16763-3_7.
58. Heyszl J, Ibing A, Mangard S, Santis FD, Sigl G. Clustering algorithms for non-profiled single-execution attacks on exponentiations. In: Smart card research and advanced applications—12th international conference, CARDIS 2013. Berlin, Germany, November 27–29, 2013. Revised Selected papers. 2013. p. 79–93. doi:10.1007/978-3-319-08302-5_6. http://dx.doi.org/10.1007/978-3-319-08302-5_6.
59. Perin G, Imbert L, Torres L, Maurine P. Attacking randomized exponentiations using unsupervised learning. In: Constructive side-channel analysis and secure design—5th international workshop, COSADE 2014, Paris, France, April 13–15, 2014. Revised selected papers, 2014. p. 144–60. doi:10.1007/978-3-319-10175-0_11. http://dx.doi.org/10.1007/978-3-319-10175-0_11.
60. Lerman L, Poussier R, Bontempi G, Markowitch O, Standaert F-X. Template attacks vs. machine learning revisited (and the curse of dimensionality in side-channel analysis). In: Mangard S, Poschmann AY, editors. Constructive side-channel analysis and secure design. Lecture notes in computer science (LNCS). Springer; 2015. p. 20–33.
61. Specht R, Heyszl J, Kleinsteuber M, Sigl G. Improving non-profiled attacks on exponentiations based on clustering and extracting leakage from multi-channel high-resolution EM measurements. In: Constructive side-channel analysis and secure design—6th international workshop, COSADE 2015, Berlin, Germany, April 13–14, 2015. Revised selected papers, 2015. p. 3–19. doi:10.1007/978-3-319-21476-4_1. http://dx.doi.org/10.1007/978-3-319-21476-4_1.
62. Özgen E, Papachristodoulou L, Batina L. Classifcation algorithms for template matching. In: IEEE international symposium on hardware oriented security and trust, HOST 2016, McLean, VA, USA; 2016 (to appear).
63. Bramer M. Chapter 3, introduction to classifcation: naïve bayes and nearest neighbour. In: Principles of data mining. undergraduate topics in computer science. London: Springer; 2013. p. 21–37. ISBN: 978-1-4471-4883-8. doi:10.1007/978-1-4471-4884-5_3. http://dx.doi.org/10.1007/978-1-4471-4884-5_3.
64. Alpaydin E. Chapter 13, kernel machines. In: Introduction to machine learning.
65. Batina L, Chmielewski L, Papachristodoulou L, Schwabe P, Tunstall M. Online template attacks. In: Proceedings progress in cryptology—INDOCRYPT 2014—15th international conference on cryptology in India, New Delhi, India, December 14–17, 2014. p. 21–36.
66. Dugardin M, Papachristodoulou L, Najm Z, Batina L, Danger J, Guilley S. Dismantling real-world ECC with horizontal and vertical template attacks. In: Constructive side-channel analysis and secure design—7th international workshop, COSADE 2016, Graz, Austria, April 14–15, 2016 (to appear).

67. Corporation A. ATMEL AVR32UC technical reference manual. ARM Doc Rev.32002F, 2010. http://www.atmel.com/images/doc32002.pdf.
68. Hutter M, Schwabe P. NaCl on 8-bit AVR microcontrollers. In: Youssef A, Nitaj A, editors. Progress in cryptology AFRICACRYPT 2013, vol. 7918. LNCS, Springer; 2013. p. 156–72.
69. Bernstein DJ, Duif N, Lange T, Schwabe P, Yang BY. High-speed high-security signatures. In: Preneel B, Takagi T, editors. Cryptographic hardware and embedded systems CHES 2011, vol. 6917. LNCS. see also full version [14]. Springer; 2011, p. 124–42.
70. Bernstein DJ, Duif N, Lange T, Schwabe P, Yang B-Y. High-speed high-security signatures. J Crypt Eng. 2012;2(2):77–89. http://cryptojedi.org/papers/#ed25519, see also short version [13].
71. Hisil H, Wong KK-H, Carter G, Dawson E. Revisited edwards curves. In: Pieprzyk J, editor. Advances in cryptology ASIACRYPT, vol. 5350. LNCS, Springer; 2008. p. 326–43.
72. Joye M, Tymen C. Protections against differential analysis for elliptic curve cryptography. In: Proceedings of Cryptographic hardware and embedded systems—CHES 2001, third international workshop, Paris, France, May 14–16, 2001. Generators, 2001. p. 377–90. doi:10.1007/3-540-44709-1_31. http://dx.doi.org/10.1007/3-540-44709-1_31.
73. Joye M. Smart-card implementation of elliptic curve cryptography and DPA-type attacks. In: Quisquater J-J, Paradinas P, Deswarte Y, Kalam A AE, editors. Smart card research and advanced applications VI, vol. 135. IFIP international federation for information processing. Kluwer Academic Publishers, Springer; 2004. p. 115–25.

Chapter 4
Practical Session: Differential Power Analysis for Beginners

Jiří Buček, Martin Novotný and Filip Štěpánek

4.1 Introduction

Differential Power Analysis (DPA) is a powerful method for breaking the cryptographic system. The method does not attack the cipher, but the physical implementation of the cryptographic system. Therefore, even systems using modern strong ciphers like AES are vulnerable to such attacks, if proper countermeasures are not applied.

The DPA method uses the fact that every electronic system has a power consumption. If you measure the power consumption of digital system, you will probably see the power trace like in Fig. 4.1 with its peaks on rising and falling edges of clock. If the digital system runs an encryption and if you run this encryption several times using various input data, you may notice slight variations in power traces, as shown in Fig. 4.1. These variations are caused by many factors (varying temperature, etc.), but one of them are varying processed (inner) data. DPA utilizes the fact that power consumption depends on processed data (e.g., number of ones and zeros in processed byte) to break the cryptographic system.

To demonstrate the power of power analysis we prepared this tutorial for you. Before we start, please, download all necessary materials from the web. You will find the compressed archive at the address http://users.fit.cvut.cz/~novotnym/DPA.zip (the file has about 250 MB). Uncompressed materials can be found also at address http://users.fit.cvut.cz/~novotnym/DPA, which might be useful if you have problems

J. Buček · M. Novotný (✉) · F. Štěpánek
Faculty of Information Technology, Czech Technical University in Prague,
Praha, Prague, Czech Republic
e-mail: martin.novotny@fit.cvut.cz; novotnym@fit.cvut.cz

J. Buček
e-mail: jiri.bucek@fit.cvut.cz

F. Štěpánek
e-mail: filip.stepanek@fit.cvut.cz

N. Sklavos et al. (eds.), *Hardware Security and Trust*,
DOI 10.1007/978-3-319-44318-8_4

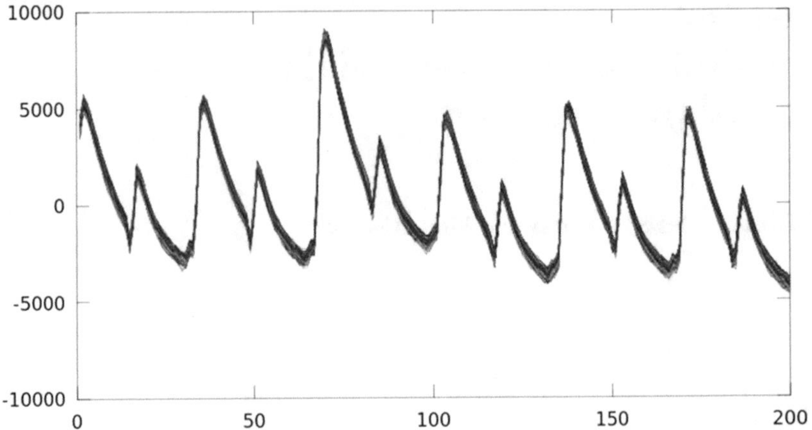

Fig. 4.1 The figure shows 500 power traces in the same time interval of 200 samples. Each power trace is run for unique input data, power traces are overlapped. Variations in power traces are caused by variations in processed data

to download the whole archive. The compressed archive contains two folders. In folder *Analysis* you will find files used in Sect. 4.2. Files used in Sect. 4.3 are available in folder *Measurement*.

You do not need to perform any measurement with an oscilloscope, as we have done these measurements for you. In folder *Analysis* you will find two sets of measurements, one set for a known key `00 11 22 33 44 55 66 77 88 99 aa bb cc dd ee ff`, and one set for an unknow key. These sets will be used in Sect. 4.2. You are also provided with sample codes in *MATLAB* that you can use in your program/script.

However, if you are equipped with an osciloscope (e.g., PicoScope), you can make your own measurement. Several advices you will find in Sect. 4.3.

4.2 Differential Power Analysis—Key Recovery

At this point you are either given or were able to measure the power consumption (traces) of the SmartCard yourself. For each power trace you have a pair of the plaintext and the encrypted ciphertext. Therefore you have all the information you need, except of the secret key. It is the goal of the differential power analysis to extract the secret key using the mentioned traces, plaintext, ciphertext, and the knowledge of the encryption algorithm by creating the hypothesis of the power consumption and correlating it to the measured traces.

4.2.1 Method

We are not going to explain the method here. If you are not familiar with the method, you may find its explanation, e.g., in the book [1], p. 119, or you will find presentation *dpa_Lisbon.pdf* in downloaded materials.

To summarize the method, you shall go through following steps:

1. Choose an intermediate value that depends on data and key
2. Measure the power traces while encrypting the data
3. Build a matrix of hypothetical intermediate values inside the cipher for all possible keys and traces
4. Using a power model, compute the matrix of hypothetical power consumption for all keys and traces
5. Statistically evaluate which key hypothesis best matches the measured power in each individual time.

The right key (part of the right key) is determined by *key hypothesis* $\rightarrow$ *intermediate value* $\rightarrow$ *consumption*, best correlating to actually measured consumption at some moment. We repeat the analysis for other parts of key, until we determine the whole key.

4.2.2 Schedule of Your Work

We reccommend you to proceed according to the following steps:

1. Plot one trace in the program you are using (MATLAB/Octave, Mathematica, etc.). Check that it is complete.
2. Plot several traces (e.g., 1st, 10th, 50th). Check the alignment of traces (they overlay correctly, triggering works).
3. Select the appropriate part of the traces (e.g., containing the first round). Read in the appropriate number of traces.
4. Depending on your measurements, you may have to perform a correction of mean values (if your measurements "wander" in voltage over time). You can do so by subtracting from each trace its mean value.
5. Recover the secret key using the DPA with correlation coefficients. The method is summarized in Sect. 4.2.1.

4.2.3 Training Sets

In folder *Analysis* you will find two sets of measurements. One set is for known key `00 11 22 33 44 55 66 77 88 99 aa bb cc dd ee ff`, the other set is for unknown key.

4.2.3.1 Training Set for Known Key `00 11 22 33 44 55 66 77 88 99 Aa Bb Cc Dd Ee Ff`

To implement and debug your program/script, we provide testing traces of 200 AES encryptions (AES 128, with 10 rounds). We encrypted 200 plaintexts (file *plaintext-00112233445566778899aabbccddeeff.txt*), obtaining 200 ciphertexts (file *ciphertext-00112233445566778899aabbccddeeff.txt*. During encryptions we measured power traces (*traces-00112233445566778899aabbccddeeff.bin*). Each trace has a length of 370 000 samples (in this case). Each sample is represented by 8 bit unsigned value (i.e., the length of the file is 370 000 bytes * 200 traces = 74 MB).

If your program/script is correct, then you should reveal the key `00 11 22 33 44 55 66 77 88 99 AA BB CC DD EE FF`.

4.2.3.2 Training Set for Unknown Key

If you are successful with the set above, you may try to recover the unknown key. We made 150 AES encryptions (AES 128, with 10 rounds).

Files *plaintext-unknown_key.txt* and *ciphertext-unknown_key.txt* contain plaintexts and corresponding ciphertexts, which were produced by an AES encryption with an unknown key. File *traces-unknown_key.bin* stores power traces recorded during encryptions of above plaintexts. File *traceLenght-unknown_key.txt* contains information on trace length, i.e., 550000 samples in this case.

You can easily check whether you found correct key. Just take any plaintext from the file *plaintext-unknown_key.txt*, encrypt it with the key you determined by the analysis, and compare the resulting ciphertext with a corresponding ciphertext from the file *ciphertext-unknown_key.txt*. If the ciphertexts match, you found the correct key.

4.2.4 Tools

We will use a system suitable for numerical calculations. *MATLAB* seems to be a system best-tailored for our needs (matrix operations with large matrices). We also can use freeware alternative *Octave*, that is compatible with *MATLAB* in its basic functions.

Mathematica is also one of alternatives. Mathematica can `Import` data in MATLAB format (*.mat*).

You may also use other computer algebraic systems—your possible experience is welcome!

4.2.4.1 MATLAB—Using the Prepared Functions

The following code samples show, how to use the prepared functions (files) to speed-up the key recovery process.

measurement.m the code template for the key recovery process
myin.m loads the content of the text files (*plaintext.txt*, *ciphertext.txt*) generated during the measurement
myload.m loads the content of the binary files (*traces.bin*) generated during the measurement
mycorr.m is used to calculate the correlation coefficient later during the recovery process

All the files are available in archive in the folder *Analysis* and should be placed into your MATLAB project directory. The following code snippets show in more detail, how to load the appropriate data using the prepared functions and are all included in the template *measurement.m*.

In case you are new to MATLAB, you can see some basic examples in the Sect. 4.2.4.2.

MATLAB code example—loading the data

```
%%%%%%%%%%%%%%%%%%%%
% LOADING the DATA %
%%%%%%%%%%%%%%%%%%%%

% modify following variables so they correspond
% your measurement setup
numberOfTraces = 200;
traceSize = 350000;

% modify the following variables to speed-up the measurement
% (this can be done later after analysing the power trace)
offset = 0;
segmentLength = 350000;
% for the beginning the segmentLength = traceSize

% columns and rows variables are used as inputs
% to the function loading the plaintext/ciphertext
columns = 16;
rows = numberOfTraces;

%%%%%%%%%%%%%%%%%%%%%%%%%
% Calling the functions %
%%%%%%%%%%%%%%%%%%%%%%%%%

% function myload processes the binary file containing the
% measured traces and stores the data in the output matrix so
% the traces (or their reduced parts) can be used for the key
% recovery process.
% Inputs:
%   'file' - name of the file containing the measured traces
```

```
%    traceSize - number of samples in each trace
%    offset - used to define different beginning of the power trace
%    segmentLength - used to define different/reduced length of the power trace
%    numberOfTraces - number of traces to be loaded
%
% To reduce the size of the trace (e.g., to speed-up the
% computation process) modify the offset and segmentLength
% inputs so the loaded parts of the traces correspond to the
% trace segment you are using for the recovery.
traces = myload('traces.bin', traceSize, offset, segmentLength, numberOfTraces);

% function myin is used to load the plaintext and ciphertext
% to the corresponding matrices.
% Inputs:
%    'file' - name of the file containing the plaintext or ciphertext
%    columns - number of columns (e.g., size of the AES data block)
%    rows - number of rows (e.g., number of measurements)
plaintext = myin('plaintext.txt', columns, rows);
ciphertext = myin('ciphertext.txt', columns, rows);

%%%%%%%%%%%%%%%%%%%%%%%%%%%%%%%%%%%%%%%%%%%%%%%%%
% EXERCISE 1 — Plotting the power trace(s): %
%%%%%%%%%%%%%%%%%%%%%%%%%%%%%%%%%%%%%%%%%%%%%%%%%
% Plot one trace (or plot the mean value of traces) and check
% that it is complete and then select the appropriate part of
% the traces (e.g., containing the first round).

% —> create the plots here <—
```

MATLAB code example—using the correlation coefficients

```
%%%%%%%%%%%%%%%%%%%%%%%%%%%%%%%%%%%%%%
% EXERCISE 2 — Key recovery: %
%%%%%%%%%%%%%%%%%%%%%%%%%%%%%%%%%%%%%%
% Create the power hypothesis for each byte of the key and then
% correlate the hypothesis with the power traces to extract the
% key.
% Task consists of the following parts:
%    - create the power hypothesis
%    - extract the key using the results of the mycorr function

% variables declaration
byteStart = 1;
byteEnd = 16;
keyCandidateStart = 0;
keyCandidateStop = 255;

% for every byte in the key do:
for BYTE=byteStart:byteEnd

    % Create the power hypothesis matrix (dimensions:
    % rows = numberOfTraces, columns = 256).
```

```
    % The number 256 represents all possible bytes (0x00..0xFF).
    powerHypothesis = zeros(numberOfTraces,256);
    for K = keyCandidateStart:keyCandidateStop
        % —> create the power hypothesis here <—
    end;

    % function mycorr returns the correlation coeficients matrix
    % calculated from the power consumption hypothesis matrix
    % powerHypothesis and the measured power traces. The
    % resulting correlation coeficients stored in the matrix CC
    % are later used to extract the correct key.
    CC = mycorr(powerHypothesis, traces);

    % —>        do proper operations here      <—
    % —> to find the correct byte of the key <—

end;
```

4.2.4.2 MATLAB for Beginners

Here you find several useful commands. We are working in certain working directory where all working files and scripts (files *.m*) are placed.

Almost all MATLAB objects are matrices. Column or row vector are special cases, however, generally we are working with n-dimensional arrays. Almost all numbers are of type double.

```
% example (this is a comment)
% matrix creation:
a = [1,2,3;4,5,6;7,8,9]
% we have defined the variable a, the result has been printed
b = rand(100,100);
% semicolon (;) suppresses printing the result (important for huge
    data)
% showing part of a matrix b:
b(1:10,5:7)
% matrix multiplication (addition/subtraction/division) works:
c = [2,0,0;0,2,0;0,0,2]
a * c
% for entry-by-entry multiplication, we use .*
a .* c
```

Vectors are special cases of matrices

```
% vectors are special cases of matrices
v = [1,3,5,7]       % row vector
v(1,:)              % equivalent to v
v(1,3:4)            % part of v
% transposition
v'                  % creates column vector
% special matrices
```

```
zeros(3,3)
ones(3,3)
eye(3,3)
rand(3,3)
% indexing by a vector
iv = [3,4,1,2]
v(iv)
% by indexing we can create originally not existing components
v
v([1,1],1:3)
v([1,1,1],:)
v(ones(1,5),:)
v'(:,ones(1,5))      % works only in Octave
```

Graph plot

```
% graph plot
e = rand(1,100);
plot(e)
f = rand(1,100);
hold on              % adding the second trace into the graph
plot(f)
% if x is a matrix:
plot(x)              % plots the set of traces by columns of x
plot(x')             % plots the set of traces by rows of x (using transposition)
```

Cycles

```
% how to write cycles
for i=1:10
  for j=1:20
    x(i)=bitxor(v(i),w(j));
  end
end
```

Manipulating files

```
% Manipulating files

% open file for reading:
MyFile = fopen ('myFile.bin', 'r');
% skip in Myfile from current position ('cof') by Offset:
fseek(MyFile, Offset, 'cof');
% read Number of uint8s to the vector Values (from the current position):
Values = fread(MyFile, Number, 'uint8');
% close MyFile:
fclose(MyFile);

% Manipulating text files

% open file for reading:
TextFile = fopen('myTextFile.txt','r');
```

```
% reading line from TextFile:
Line = fgets(TextFile );
% reading 16 values from the Line according to the pattern (like in C ):
[values , l] = sscanf(Line , '%02x ', 16);
```

Printing data in hex-form

```
% suppose the key is stored here in the key array:
key=[1,2,3,4,5,6,7,8,9,1,2,3,4,5,6,7];

% to print it in hex-form run the following for-cycle:
for i=1:16
    fprintf('Byte %d of the key is 0x%2.2X \n', i, key(i));
end;
```

4.2.4.3 MATLAB/Octave Tips

- For *xoring* of values you may use the `bitxor` function. This function also performs bit-wise xor of vectors and matrices (of the same size).
- The average value (mean value) is calculated by the function `mean`. If `a` is a matrix, then `mean(a)` is a (row) vector of mean values of columns, while `mean(a,2)` is a (column) vector of mean values of rows (which is probably what we want).
- If you like to extend (copy) the column vector into matrix, use indexing (e.g., you like to extend vector *b* into matrix having 100 columns):

  1. By indexing: `b_mat = b(:,ones(1,100));`
  2. By replication: `b_mat = repmat(b,1,100);`

- You can use arrays `SubBytes` and `byte_Hamming_weight` (see the file *tab.mat*). Remember that the first index of an array is equal to 1, therefore you probably need to increment index by 1, e.g.,: `a = SubBytes(x + 1); b = byte_Hamming_weight(a + 1);` This works also for matrices (!)—if `x` is a matrix, then `SubBytes` applies to all its elements (the result is a matrix again).

4.2.4.4 Mathematica—Tips

Mathematica can import files in MATLAB format (matrix format, *.mat*) using function `Import`. Function `Import` may be highly memory demanding, as it always imports the whole file.

```
$HistoryLength = 0; (*saving the memory*)
SetDirectory[NotebookDirectory[]]
NUMBER = 500;
t = Import["traces-part.mat"][[1]][[1 ;; NUMBER]];
```

```
{MemoryInUse[], MaxMemoryUsed[]} (*checking the occupied memory*)
ListLinePlot[t[[1 ;; 100, 1 ;; 100]]]
```

Use the function `Mean` for elimination of a systematic error of measurement (different DC component between traces). `Mean` applied to a matrix returns a vector of means of columns. However, we need means of rows, i.e. we have to use `Mean[Transpose[...]]`.

4.3 DPA—Measurement with an Oscilloscope

First we will set up a basic measurement of the smart card consumption. We will use JSmartCard Explorer to communicate with the card, and PicoScope 6 GUI to establish basic parameters of the measurement. Then, we will switch to a separate program that will control both the card and the oscilloscope and will perform a series of measurements needed for the DPA attack.

4.3.1 Preparation of the Measurement

1. Connect the card reader to your computer and insert the SmartCard into the reader, as shown in Fig. 4.2.
2. Run *JSmartCard Explorer* from Primiano Tucci [2]. (In Java. Compiled JAR file you find either on web [3] or in file *JSmartCardExplorer.jar* in downloaded archive).

   - Press the *Connect* button to connect to the SmartCard. (Status should be green.)
   - Fill-in the fields *Class* (80), *INS* (60), *P1* (00), *P2* (00), *Data IN*, and *Le* (10) as shown in Fig. 4.3 (all in HEX).
   - Press *Send*. The card should run AES encryption over the entered data and return the ciphertext, as shown in Fig. 4.3.

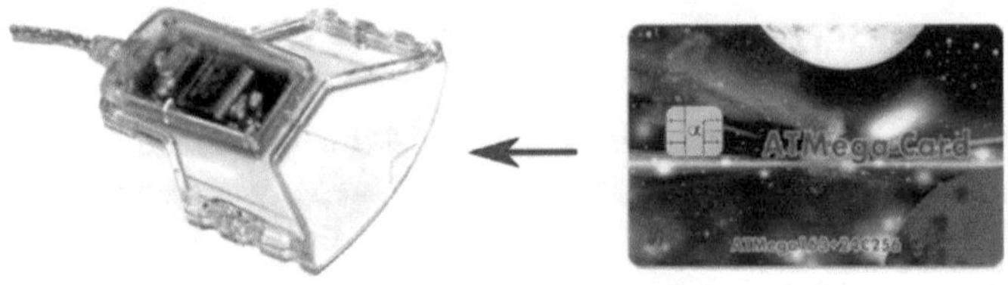

Fig. 4.2 Connect the card reader to your computer and insert the SmartCard into the reader

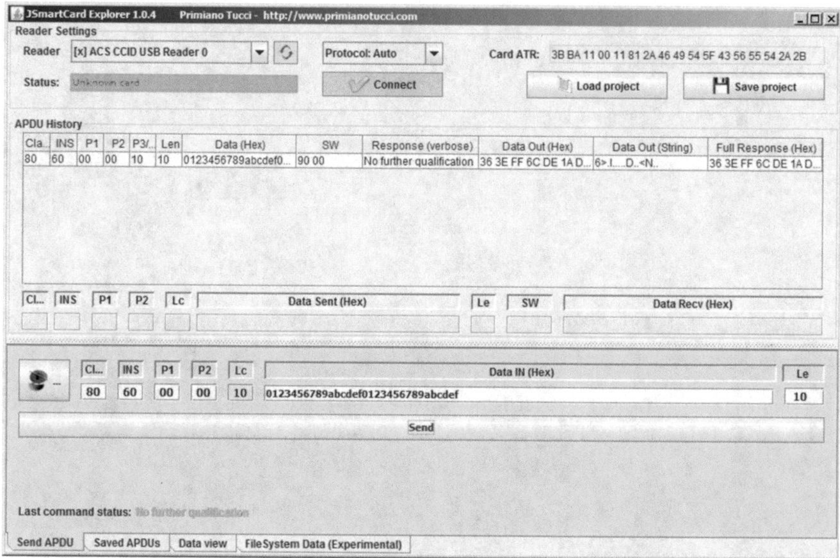

Fig. 4.3 Fill-in proper fields and press *Send*

Fig. 4.4 Insert card into the measuring adaptor (*green PCB*), then insert the measuring adaptor into the reader

3. To measure the card power consumption, we will be using Picoscope 5204 (and 5203) and two oscilloscope probes. Connect the probes to channels A (blue, trigger), and B (red, trace measurement).
4. Remove the card from the reader and insert it into the measuring adaptor (green PCB), then insert the measuring adaptor into the reader. See Fig. 4.4
5. Connect the Picoscope probes to the measuring adaptor. Unlike to Fig. 4.5, set the trigger probe (channel A, blue) to the X10 position and the measurement probe (channel B, red) to the X1 position.
6. Connect the Picoscope to a free USB port of your computer (if not already connected).
7. For the measurement you need the following software:

   - PicoScope 6 software with drivers. You can download it from [4]. You can find it also in downloaded materials as a file *PicoScope6_r6_8_11.exe*.

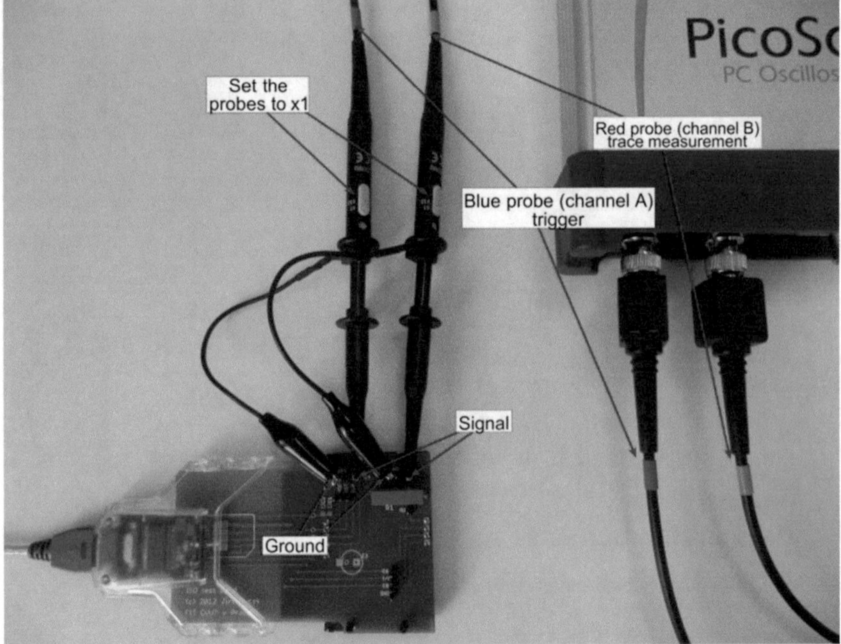

Fig. 4.5 Connect the Picoscope probes to the measuring adaptor. Set the trigger probe (channel A, *blue*) to the X10 position and the measurement probe (channel B, *red*) to the X1 position

- Software Development Kit. The relevant files should be included in the Visual Studio project below. If not, you can download the SDK from the web [5] or you may find it in downloaded materials as a file *PS5000sdk_r10_5_0_32.zip*.
- Library for working with smart cards (*WINSCard.lib*). This should be included in the installation of Visual Studio. (It is a part of Microsoft Windows SDK.)

8. Run the PicoScope 6 program
9. You should make the following settings:

   - *Timebase*: 500us/Div, x1 (zoom), 1 MS (samples)
   - *Channel A*: +-1V DC
   - *Channel B*: +-1V DC
   - *Trigger*: Auto (after tuning the settings, switch to Repeat)
   - *Trigger Event*: Simple Edge, Rising
   - *Trigger Channel*: A
   - *Trigger Threshold*: 200 mV

   Warning: These settings may need to be adjusted according to the particular card and other circumstances.
10. Set the *Single* measurement at the oscilloscope, and send data to the card using *JSmartCard Explorer* from Primiano Tucci. You should see a waveform like in Fig. 4.6.

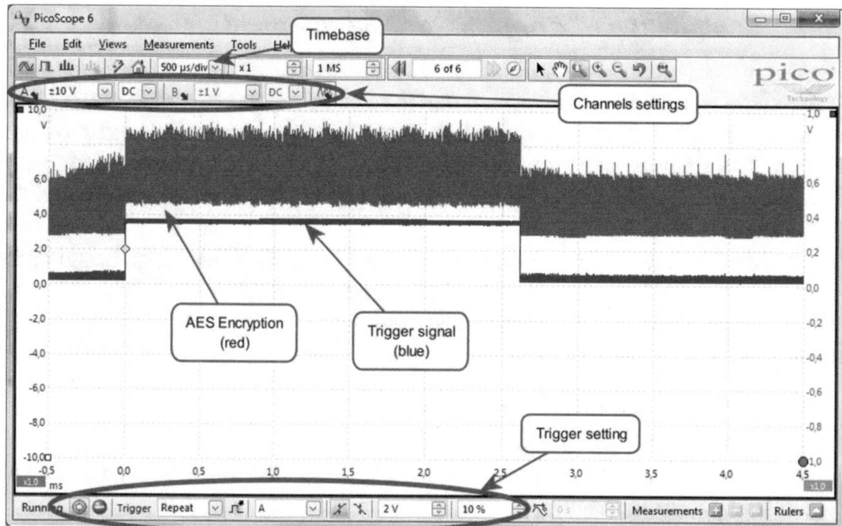

Fig. 4.6 Powertrace of one encryption

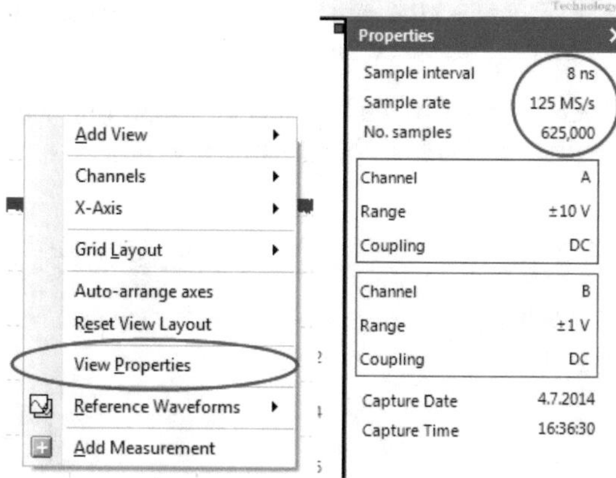

Fig. 4.7 Menu Properties of PicoScope program

11. Display the Properties panel by right-clicking somewhere in the window and selecting View Properties (see Fig. 4.7).
12. From the Properties panel remember the following values:

    - Sample interval,
    - Sample rate,
    - No. samples.

    We will need these values later, when setting the measurement program.

4.3.2 Compilation of Program for Measurement

At this stage, you should have verified that the SmartCard works correctly (responds to the command for AES encryption), and that the signals from the card look reasonable. Press *Disconnect* or close *JSmartCard Explorer*, and close *PicoScope 6 GUI*. We will use a separate program to control both the SmartCard reader and the oscilloscope.

For measurement it is necessary to adjust and compile C++ program stored in an archive *Pico5000.zip*. Zip file contains source files and Microsoft Visual Studio project. After extracting the archive and opening the project in Microsoft Visual Studio you have to check the following settings in project properties:

1. Paths to include and library directories, see Fig. 4.8.
2. Paths to additional dependencies, see Fig. 4.9.
3. In source file *main.cpp* set up the measuring channels, trigger voltage level, and number of measurements.

Compile the program (*Build* → *Build Solution*). Before running the program do not forget:

- to disconnect the card in JSmartCard Explorer and
- quit the PicoScope program,

otherwise the card and/or the PicoScope would be occupied, hence the measuring program will not be able to connect to it.

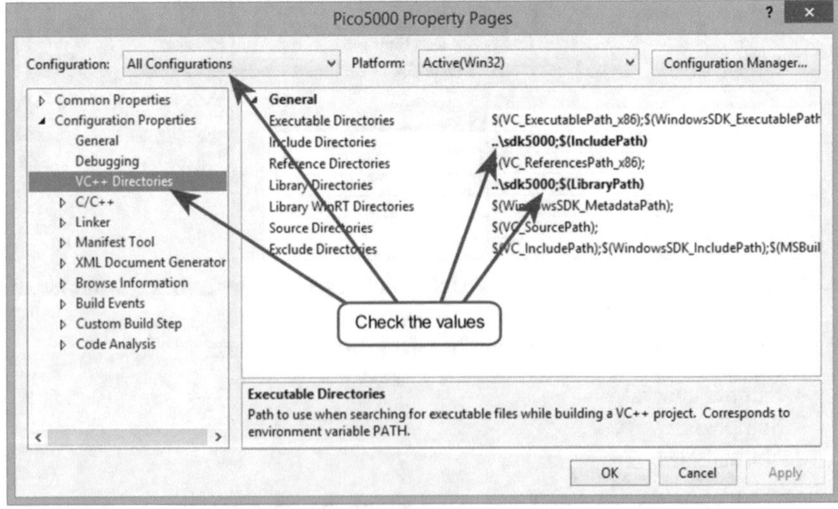

Fig. 4.8 Visual Studio project setup—include and library directory paths

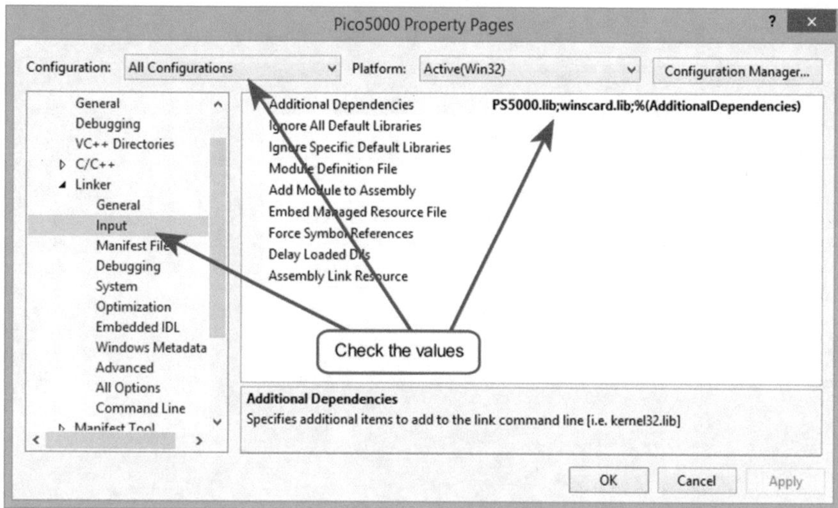

Fig. 4.9 Visual Studio project setup—additional dependencies

Measured data are in file *traces.bin*, plaintext and cipher text in files *plaintext.txt* and *ciphertext.txt* and length of one measurement is stored in file *traceLength.txt*.

Now you have measured data to be used for DPA.

References

1. Mangard S, Oswald E, Popp T. Power analysis attacks: revealing the secrets of smart cards. US: Springer; 2008.
2. Tucci P. JSmartCardExplorer. https://www.primianotucci.com/os/smartcard-explorer. Accessed 02 Mar 2016.
3. Tucci P. JSmartCardExplorer. http://downloads.sourceforge.net/jsmart-card/JSmartCardExplorer.jar. Accessed 02 Mar 2016.
4. Picotech. PicoScope 6 software with drivers. http://downloads.picotech.com/winxp/PicoScope6_r6_8_11.exe. Accessed 02 Mar 2016.
5. Picotech. Software development kit. http://dl.picotech.com/drivers/PS5000sdk_r10_5_0_32.zip. Accessed 02 Mar 2016.

Chapter 5
Fault and Power Analysis Attack Protection Techniques for Standardized Public Key Cryptosystems

Apostolos P. Fournaris

5.1 Introduction

The basic building block of any security protocol is its cryptographic algorithms and their primitive operations. While there exist many cryptographic algorithms only few of them are standardized. In several of them, their cryptographic primitives have considerable similarities. This is especially true in public key cryptography where the real, day-to-day, security scene is dominated by security products relying on public key cryptography schemes that are based on RSA, El Gamal or ECC approaches. So, considerable research is focused on enhancing the security of such standardized schemes implementations without reducing those implementation performance.

The mathematical backbone of RSA, El-Gamal and ECCs is the integer factorization problem (IFP), the discrete logarithm (DLP) and elliptic curve discrete logarithm problem (ECDLP) respectively. Those problems rely on the arithmetic operations of modular exponentiation (ME) and scalar multiplication (SM) that from number theoretic perspective are very closely related. RSA and El-Gamal is structured around $\mathbb{Z}_n^*$ multiplicative and additive cyclic group, i.e., where addition and multiplication operations are defined. Modular exponentiation in $\mathbb{Z}_n^*$ for RSA is defined as $c^e \bmod n$ where c is an RSA message, e is the exponent (public or private RSA key) and n is the public modulus. To reduce high bit length of involved numbers needed to keep the IFP or DLP hard, we can replace $\mathbb{Z}_n^*$ with a different abelian group where these problems are harder. Such group is the Elliptic Curve group $E(F)$ where F is a finite field. However, since this group is additive, all $\mathbb{Z}_n^*$ multiplication operations are replaced by their additive equivalent. Thus, in ECC schemes, all Elliptic Curve points $P : (x, y)$ are defined over the additive group $E(F)$ where F is a finite field in which each EC point's coordinates (x, y) belong to F and instead of $\mathbb{Z}_n^*$ multiplication, addition between EC points is performed while instead of $\mathbb{Z}_n^*$ squaring EC

A.P. Fournaris (✉)
Computer Informatics Engineering Department, Technological Educational Institute of Western, Antirion, Greece
e-mail: apofour@ieee.org; afournaris@teimes.gr

N. Sklavos et al. (eds.), *Hardware Security and Trust*,
DOI 10.1007/978-3-319-44318-8_5

doubling (as a special case of addition) is performed. $\mathbb{Z}_n^*$ ME can be realized as series of modular multiplications and squaring operations. Similarly, due to the equivalence of $\mathbb{Z}_n^*$ multiplication to $E(F)$ EC point addition, $E(F)$ scalar multiplication (SM), defined as $e \cdot P$ where $P \in E(F)$ (P is denoted also as base point) and $e \in F$, from engineering perspective, is equivalent to $\mathbb{Z}_n^*$ ME and can be realized as a series of $E(F)$ EC point addition and $E(F)$ point doubling operations.

Regardless of the fact that $\mathbb{Z}_n^*$ operations and F operations (i.e., operation between EC point coordinates) can have significant differences (e.g., when F is a binary extension field), the $\mathbb{Z}_n^*$ multiplication to $EC(F)$ addition equivalence hints that the two public key cryptographic primitives design (ME and SM) follow similar principles for achieving efficiency in hardware terms (chip covered area, resources, computation time, power consumption). Traditional ME and SM algorithms like multiply and square or double-and-add algorithms respectively are very close in concept.

Similar ME and SM design principles, however, lead to common implementation attack techniques and approaches. Implementation attacks target an actual implementation of a cryptographic algorithm and exploit information leakage (side channel attack) or faulty behavior (fault injection) of the implementation's physical characteristics (power dissipation, timing, electromagnetic emission, etc.). As expected, side channel (SCA) and fault analysis (FA) attacks in ME or SM designs require similar SCA-FA countermeasures. In this book chapter, apart from FAs, our research interest is focused on SCAs relying on power dissipation, known as power analysis attacks (PA) but the proposed countermeasures can be also applicable to other SCAs (e.g., relying on electromagnetic emission or timing).

In this book chapter, expanding the work of [1], the concept of a unified SCA-FA protection mechanism both for ME and SM is explored. This mechanism is capable of thwarting a wide range of existing PA and FA attack approaches. The proposed approach is a variation of the Montgomery Power Ladder algorithm for ME/SM that is sufficiently modified in order to counter "vertical" and "horizontal" simple and advanced SCAs (focusing on PAs). To achieve that goal, the randomization technique is adopted in the proposed algorithm by introducing a random element $\in\mathbb{Z}_n^*$ or $E(F)$ along with the message/base point in every algorithmic round. This randomization is propagated and extended in each round and is only removed after the last round of the proposed algorithm. The high regularity of the Montgomery Power Ladder algorithm and its intrinsic parallelism provide high performance as well as additional resistance against SCAs-PAs. The proposed algorithm takes advantage of the intrinsic mathematical coherence between intermediate algorithmic values, offered by Montgomery Power Ladder, to detect possible faults (following the infective computation principle) thus providing FA resistance. Attempts to bypass successfully the fault detection mechanism by injecting a second fault lead to non-usable information by an attacker since the ME/SM result is released (unblinded) only after passing the fault detection check. The above countermeasures are combined in an harmonic way so that they do not introduce new vulnerabilities.

The rest of the paper is organized as follows. In Sect. 5.2 existing SCA-PA and FA approaches and countermeasures on ECC and RSA systems are presented. In Sect. 5.3 the proposed approach is described and a security analysis of the algorithm is made in Sect. 5.4, while Sect. 5.5 concludes the paper.

5.2 Public Key Primitive Fault and Power Attacks and Countermeasures

5.2.1 Side Channel Attacks and Countermeasures

To model side channel attacks we can adopt the approach described in [2, 3]. Assume that we have a computation C (which can be an RSA modular exponentiation or EC scalar multiplication) that consists of series of O_0 or O_1 operations that require inputs X_0 and X_1, respectively (thus $O_i(X_i)$ for $i \in \{0, 1\}$). During processing of the C computation, each operation can be linked to an information leakage variable L_i. A side channel analysis attack is possible if there is some secret information s that is shared between O_i and its leakage L_i. The ultimate goal of a side channel analysis is, by using a strategy, to deduce s from the information leakage L_i. The simplest way to achieve that is by examining a sequence of O_i operations in time to discover s. Simple SCAs (SSCAs) can be easily mounted in square-and-multiply/double-and-add algorithms used in ME/SM and are typically horizontal type of attacks meaning that they are mounted using a single leakage trace that is processed in time. When SSCAs are not possible, advanced SCAs (ASCAs) must be mounted to a ME/SM architecture to extract s.

Advanced SCAs do not focus only on the operations (eg. O_i) but also on the computation operands [3]. Advanced SCAs are focused on a subset of the calculation C (and/or O_i) and through collection of sufficiently large number N of leakage traces $L_i(t)$ for all $t \in \{1, \ldots, N\}$ using inputs $X_i(t)$ exploit the statistical dependency between the calculation on C for all X_i and the secret s. ASCAs follow the hypothesis test principle [2, 4] where a series of hypothesis $\acute{s}$ on s (usually on some bit j of s, i.e., $\acute{s}_j = 0$ or 1) is made and a series of leakage prediction values are found based on each of these hypothesis using an appropriate prediction model. The values of each hypothesis are evaluated against all actual leakage traces using an appropriate distinguisher δ for all inputs X_i so as to decide which hypothesis is correct.

SSCAs and ASCAs can follow one of two different leakage collection and analysis strategies, as originally described in [2], the vertical or horizontal approach. In the vertical approach, the implementation is used N times using either the same or different inputs each time t in order to collect traces-observations $L_i(t)$. Each observation is associated with t-th execution of the implementation. In the horizontal approach, leakage traces-observations are collected from a single execution of the implementation under attack and each trace corresponds to a different time period within the time frame of this execution. As expected, in horizontal attacks the implementation input is always the same.

Many SSCAs fit on the horizontal analysis strategy, as long as they are based on a single implementation execution leakage collection. Such attacks enable the attacker to discriminate O_1: modular multiplication (RSA) or point addition (ECC) from O_0: Modular squaring (RSA) or point doubling (ECC) in time thus revealing all bits of the secret s (the exponent (RSA) or secret scalar (ECC)). There also exist ASCA horizontal attacks that take advantage of the fact that each O_i operation when implemented in an existing generic processor, is broken into a series of digit based operations (e.g., word-based field multiplications) that are all associated to the same bit of the secret exponent/scalar. Such attacks are the Big Mac attack [5], the Horizontal Correlation Analysis attack (HCA) [6] or the Horizontal Collision Correlation attack (HCCA) [2, 3] that are described both for RSA and ECC designs.

There is a very broad range of vertical approach-based attacks on ME/SM implementations including sophisticated SSCAs and most of the ASCAs. Such SSCAs that require more than one ME/SM executions (e.g., two executions) include comparative SCAs (originally focused on Power attacks (PAs)) like the doubling attack (collision based attack) [7] (DA attack) and its variant, relative doubling attack (RDA attack) [8] or the chosen plaintext attack in [9] (also known as 2-Torsion Attack (2-TorA) for ECC). Vertical SSCA include also attacks applied specifically to SM, like the refined PA (RPA) or zero PA (ZPA) where a special point P_0 (that can zero a P coordinate) is fed to an SM accelerator, thus enabling scalar bit l recovery through a vulnerability at round l.

Most ASCAs follow the vertical attack paradigm. Their success rate is associated with the number of traces that are needed to be processed vertically in order to reveal the secret s. The most widely used ASCA vertical attack is Differential Attack (DSCA) originally proposed by Kocher in [10] that is later expanded into the more sophisticated Correlation SCA (requiring less traces to reveal the secret than DSCA) [11] and collision correlation attack [12–14] that can be mounted even if the attacker does not have full control of the implementation inputs.

Recently, researchers have shown that appropriate combination of vertical and horizontal attacks can enhance SCA success rate even against implementations that have strong SCA countermeasures [14, 15]. These publications are mainly based on vertical attacks that use horizontal attacks to bypass randomization/blinding countermeasures.

Countermeasures: SSCAs are thwarted by making the leakage trace of O_1 indistinguishable from the leakage trace of O_0. This can be achieved by more sophisticated (regular) ME/SE algorithms, like the square-and-multiply always/ double-and-add always technique or the Montgomery power ladder (MPL) technique [16] (presented in the following Table) or by applying the atomicity principle in the existing square-and-multiply / double-and-add ME/SM algorithm. Atomicity is realized by braking each O_i operation into atomic blocks (e.g., the same field operations) that are arranged in such way in time that they follow the same sequence for both O_1 and O_0. On the other hand, regular ME/SM algorithms provide SSCA resistance by making

MPL for RSA primitives
Input: $c, e = (1, e_{t-2}, ...e_0) \in \mathbb{Z}_n^*$
where n is the public modulus
Output: $S = c^e \ mod \ n$
Initialization: $T_0 = 1, T_1 = c$
For $i = t - 1$ **to** 0
  If $e_i = 1$ **then**
   $T_1 = T_1^2 \ mod \ n$
   $T_0 = T_0 \cdot T_1 \ mod \ n$
  else
   $T_0 = T_0^2 \ mod \ n$
   $T_1 = T_0 \cdot T_1 \ mod \ n$
Return: T_0

MPL for ECC primitives
Input: $P, \in E(F), e = (e_{t-1}, e_{t-2}, ...e_0) \in F$
Output: $S = (e \cdot P)$
Initialization: $T_0 = O, T_1 = P$
For $i = t - 1$ **to** 0
  If $e_i = 1$ **then**
   $T_1 = 2 \cdot T_1$
   $T_0 = T_0 + T_1$
  else
   $T_0 = 2 \cdot T_0$
   $T_1 = T_0 + T_1$
Return: T_0

the number of O_i operations constant in each ME/SM round (that processes one bit of the secret exponent/scalar). Unfortunately, the above countermeasures can be bypassed when each O_i operation is realized by $\mathbb{Z}_n^*$ operations or F operations (for ME or SM respectively) that are implemented as a series of word-based operations (typical case in software implementations). In such case, horizontal attacks like the Big Mac, HCA, HCCA are still successful. Furthermore, the above countermeasures are thwarted by all vertical type of attacks including DA, RDA, and 2-Torsion and all ASCAs.

Randomization is a favorable solution for countering ASCAs (both horizontal and vertical). Using randomization, the sensitive information (exponent or scalar) is disassociated from the leakage trace and is hidden by multiplicatively or additively blinding this information using a random Group ($\mathbb{Z}_n^*$ or $E(F)$) element. This hiding/blinding involves exponent, public modulus or message multiplication with a random number in the RSA case, or adding a random R point to the SM base point P, multiplying with a random element of F the base point's projective coordinates as well as applying EC or finite field random isomorphisms (Coron's Countermeasures [17]). Many of the above countermeasures do not fully protect an ME/ SM architecture from CSCA, CCSCA (and the SM specific attacks of RPA, ZPA [18]). This is more evident in ECC SM implementations where attackers have managed to defeat all 3 of Coron's countermeasures (for some regular SM algorithms). For example, in SSCA resistant algorithms, like the BRIP method [19] (presented below) where the same random number is added to each round's point values (thus creating a vulnerability [20]), randomization (base point blinding) may not prevent RDA or 2-TorA. Researchers have also shown that blinding cannot protect an ME/SM implementation if $\mathbb{Z}_n^*$ operations or F operations (for ME or SM respectively) are implemented as a series of word-based operations. In such case, horizontal attacks (HCA, HCCA) or vertical-horizontal attack combinations are successful in revealing the secret s [14, 15]. Yet still, message/base point blinding can resist horizontal attacks as long as the bit length of the employed random element is large enough [6].

BRIP for RSA primitives
Input: $c, B, B^{-1}, e = (1, e_{t-2}, ...e_0) \in \mathbb{Z}_n^*$ where n is the public modulus
Output: $S = c^e \ mod \ n$
Initialization: $T_0 = B, T_1 = B^{-1}$, $T_2 = c \cdot B^{-1} \ mod \ n$
For $i = t - 1$ **to** 0
  $T_0 = T_0^2 \ mod \ n$
  If $e_i = 1$ **then**
    $T_0 = T_0 \cdot T_2 \ mod \ n$
  else
    $T_0 = T_0 \cdot T_1 \ mod \ n$
Return: $T_0 = T_0 \cdot T_1 \ mod \ n$

BRIP for ECC primitives
Input: $P, B, \in E(F), e = (e_{t-1}, e_{t-2}, ...e_0) \in F$
Output: $S = e \cdot P$
Initialization: $T_0 = B, T_1 = -B$, $T_2 = P - B$
For $i = t - 1$ **to** 0
  $T_0 = 2 \cdot T_0$
  If $e_i = 1$ **then**
    $T_0 = T_0 + T_2$
  else
    $T_0 = T_0 + T_1$
Return: $T_0 = T_0 + T_1$

5.2.2 Fault Attack and Countermeasures

Fault attacks can be injected in various parts of the RSA/ECC implementation including storage elements, control instructions or computation units as a whole. Bellcore researchers introducing FAs in public key systems, have shown that RSA, especially CRT[1] RSA, is very vulnerable against fault attacks [21]. Similarly, FAs have been very successful in ECC implementations. There exist various FAs aiming the SM implementation, like C and M safe error attacks where the value of a single bit of the scalar e is changed and it is observed if this action leads to a different point multiplication outcome or not (safe error). There also exist FAs focusing on a weak curve-based fault analysis including invalid base point attacks where by injecting a fault in the SM base point, this point with high probability becomes a point of a weak curve.[2] This approach can be expanded into invalid curve attacks, where any unknown fault in any part of the hardware implementation (memory, buses, registers etc.) influencing any EC parameter can possibly lead to a transition to a weak curve [22]. By specializing the fault injection process to the x EC point coordinate (as long as the y coordinate is not used), more promising attack results can be provided by transferring SM calculations to a weak twist of the original EC with high probability (twist curve FAs) [23].

Apart from the patented approach of Shamir [24] (Shamir's trick), early attempts to thwart the Bellcore attack and EC SM fault attacks were based on infective computation [25]. Through this approach, any computational errors introduced by a fault will propagate throughout the computation, "infecting" all intermediate variable thus ensuring that the final result always becomes faulty and appears random and useless to the adversary in the end. After an initial attempt on this concept by Yen in [26], in the case of RSA, insecurities were found by Blömer et al. [27], thus the infective computing approach was enhanced with a fault detection mechanism based on the introduction of public modulus (n) multiplicative masking (BOS scheme).

[1]Chinese Remainder Theorem.

[2]A weak elliptic curve is a curve that can be cryptanalyzed easily.

BOS scheme was insecure in several possible thread models [28], as shown in [29, 30]. More than one fault can be carefully injected, as shown by Kim and Quisquater in [31], in certain parts of the CRT and non-CRT RSA to bypass the fault detection operation as a whole; thus revealing the public modulus or its private factors (KQ scheme). This attack consists of injecting two faults, one during exponentiation and another during fault detection. To prevent such attack, the RSA outcome should be revealed and stored only after fault detection. This attack of more than one fault injections can also be applied to ECC designs to bypass the fault detection mechanism.

In the case of ECCs, similar countermeasure steps where introduced by researchers, including infective computation and fault detection [32]. However, to thwart the transition to weak ECs due to fault injection additional countermeasures could be taken into account, including point validation and EC integrity checks for invalid point and invalid curve (EC parameter) attacks. In general, the fault detection mechanism for both RSA and ECC schemes is focused on a coherency check between intermediate values during ME (RSA) or SM (ECC). This check is usually a mathematical connection between those intermediate values that is retained throughout the computation flow and is disrupted when an fault is injected. A coherency sensitive mechanism can check if the mathematical connection between those values exists or not, thus detecting an attack [33–35].

RSA and ECC implementations are very susceptible to SCA and especially power attacks (PA) especially when such attacks are combined with Fault attacks [36, 37]. Providing protection for FA or PA independently can thwart only one kind of hardware attack while adversaries usually apply a combination of different attack techniques to compromise an RSA/ECC hardware architecture. Combining more than one type of countermeasure as well as adopting and combining well-established resistance principles in an RSA/ECC implementation can achieve long-term SCA-FA resistance against such attacks [33, 34, 38]. However, combining FA and PA resistance approaches may introduce new vulnerabilities that can be exploited to attack the public key implementation system [16, 19, 37, 39, 40] thus reducing the RSA/ECC implementation overall physical attack resistance.

5.3 Proposed Approach

The broad variety and heterogeneity of PA and FA attacks implies that it is hard to design countermeasures capable of providing wide scale protection. This is further supported by the fact that PA and FA combinations apart from eliminating vulnerabilities may introduce new ones. Apart from specific design oriented countermeasures like dual rail logic and power balancing [41, 42] that must be fine-tuned to a single implementation in order to be effective, algorithmic-based countermeasures may offer a more generic protection approach that can be applied to a wide range of RSA/ECC implementations regardless of the architecture those implementations follow. Our goal is to describe such algorithmic approaches for PA and FA resistance that combine effectively different PA and FA countermeasures and offer long-term PA-FA resistance against known attacks. This research approach focus point is on well-established PA-FA resistance principles rather than specific resistance countermeasures on ME and SM accelerator units.

As a basis of the proposed algorithm approach on PA-FA resistant ME/SM, the MPL algorithm is used. The MPL algorithm is resistant against many of the mentioned attack in Sect. 5.2.1, it does not rely on dummy operations in order to hide the computation flow during ME/SM execution (modular multiplication or squaring for ME or point addition and doubling for SM) and also favors operation parallelism thus leading to fast implementations. The original MPL algorithm though offers SSCA resistance (and more specifically Simple PA resistance) and under some restrictions is horizontal attack resistant. To further enhance the MPL with ASCA resistance, we must introduce some blinding technique through additive or multiplicative randomization. Such countermeasure follows the protection technique of message/base point blinding, since it constitutes an approach that under careful application in the MPL algorithm cannot be bypassed or introduce considerable performance overhead to a ME/SM implementation. Other techniques like exponent/scalar blinding are not very efficiently implemented and are found to have vulnerabilities [36, 37]. However, message/base point blinding must be realized in such a way that it should not suffer from vulnerabilities similar to the BRIP method [20].

Assuming that all operations in the proposed algorithm are defined in a group $\mathbb{G}$, where $\mathbb{G}$ is either the multiplicative group $\mathbb{Z}_n^*$ (for RSA) or the additive group $E(F)$ (for ECC), we introduce a random element $B \in \mathbb{G}$ and its inverse $B^{-1} \in \mathbb{G}$ into the MPL computation flow that can blind the message multiplicatively ($B \cdot c \ mod \ n$, i.e., message blinding for RSA) or the base point P additively ($B + P$, i.e., base point blinding). In contrast to similar approaches, where in each ME/SM round the round's computed values are blinded with the same random element, in the proposed approach, a round's values are randomized with a different number in each round (a multiple of the random element B).

Concerning FA resistance, our approach adopts a combination of the infective computation and fault detection resistance principles, following the intermediate values mathematical coherence characteristic of the MPL algorithm. As observed in [16] and by Giraud in [33], the T_0 and T_1 value in an MPL round always satisfy the equation $T_0 = c \cdot T_1 \ mod \ n$ or $T_0 = P + T_1$ for ME or SM, respectively. Injecting a fault during computation in a T_1 or T_0 variable will ruin this coherence and by introducing an MPL coherence detection mechanism in the end of the MPL algorithm, this fault will always be detected. Finally, efficiency of the proposed approach is achieved by employing Montgomery modular multiplication for ME and by exploiting the intrinsic parallelism that exist in the MPL algorithm. The proposed PA-FA resistant algorithm is presented below in two formulations, ME for RSA and SM for ECC schemes.

FA-PA Montgomery ME algorithm for RSA primitives
Input: $c, B, B^{-1}, e = (1, e_{t-2}, \dots e_0) \in \mathbb{Z}_n^*$ where n is the public modulus
Output: $(s_0, s_1, s_2, s_4) = (B^e \cdot c^e \bmod n, B^{\overline{e}+1} \cdot c^{\overline{e}+1} \bmod n, B^{2^t} \cdot c^{2^t} \bmod n, B^{-e} \bmod n)$

Initialization: T = R^2 mod n, $s_0 = s_1 = b_R = B \cdot R$ mod n, $s_3 = s_4 = s_5 = b_{R_{-1}} = B^{-1} \cdot R$ mod n, where $R = 2^{j+2}$

1. $T_R = T \cdot c \cdot R^{-1}$ mod n
2. $s_2 = b_R \cdot T_R \cdot R^{-1}$ mod n
3. **For** $i = 0$ **to** $t - 1$

   (a) **If** $e_i = 1$ **then**
   $s_0 = s_0 \cdot s_2 \cdot R^{-1}$ mod n,
   $s_4 = s_4 \cdot s_3 \cdot R^{-1}$ mod n
   else
   $s_1 = s_1 \cdot s_2 \cdot R^{-1}$ mod n,
   $s_5 = s_5 \cdot s_3 \cdot R^{-1}$ mod n
   (b) $s_2 = s_2^2 \cdot R^{-1}$ mod n, $s_3 = s_3^2 \cdot R^{-1}$ mod n

4. $s_0 = s_0 \cdot b^{-1} \cdot R^{-1}$ mod n, $s_1 = s_1 \cdot c \cdot R^{-1}$ mod n
   $s_2 = s_2 \cdot 1 \cdot R^{-1}$ mod n, $s_4 = s_4 \cdot b \cdot R^{-1}$ mod n
5. **If** (values of i, e are not modified and $s_0 \cdot s_1 \cdot R^{-1}$ mod $n = s_2 \cdot 1 \cdot R^{-1}$ mod n) **then** return s_0, s_1, s_2, s_4 **else** return error

The above algorithm can be used for non CRT RSA or as a building block for CRT RSA primitive. It employs as inputs the message c, the random number B and its multiplicative inverse B^{-1}, the public modulus n and the exponent e. Note that e_i corresponds to the i-th bit of e and that j is the bit length of the modulus n. We assume that the multiplicative inverse of B exists, meaning that $gcd(B, n) = 1$ (B and n are relatively prime). Possible fault injection attack can be detected by checking $s_0 \cdot s_1 \cdot R^{-1} modn \stackrel{?}{=} s_2 \cdot R^{-1}$ mod n ($\mathbb{Z}_n^*$ MPL coherency check). If no fault is injected, the above equation is always true.[3] The exponentiation result can be found after fault detection by performing $s_0 \cdot s_4$ mod $n = B^e \cdot c^e \cdot B^{-e}$ mod $n = c^e$ mod n.

FA-PA SM algorithm for ECC primitives
Input: $P, B, B^{-1} \in E(F), e = (1, e_{t-2}, ...e_0) \in F$
Output: $(S_0, S_1, S_2, S_4) = (e \cdot (B + P), (\overline{e} + 1) \cdot (B + P), 2^t \cdot (B + P), (-e) \cdot B)$
Initialization: $S_0 = S_1 = B, s_3 = s_4 = s_5 = -B$

1. $S_2 = B + P$
2. **For** $i = 0$ **to** $t - 1$

   (a) **If** $e_i = 1$ **then**
   $S_0 = S_0 + S_2$,
   $S_4 = S_4 + S_3$
   else
   $S_1 = S_1 + S_2$,
   $S_5 = S_5 + S_3$
   (b) $S_2 = 2 \cdot S_2, S_3 = 2 \cdot S_3$

[3]Note that $\overline{e}$ is logical NOT of e and that $e + \overline{e} = 2^t - 1$.

3. $S_0 = S_0 - B, S_1 = S_1 + P$
   $S_4 = S_4 + B$
4. **If** (values of i, e are not modified and $S_0 + S_1 = S_2$) **then** return S_0, S_1, S_2, S_4
   else return error

The above algorithm can be applied to any EC type (Wierstrass, Hessian, Montgomery, Edwards curves etc.) under any coordinate system (affine, projective, mixed). It employs as inputs the base point P, a random point B and its additive inverse $B^{-1} = -B$, along with the scalar e. Note that e_i corresponds to the i-th bit of e and that j is the bit length of all involved finite field elements. Similar to its ME version, possible fault injection attack can be detected by evaluating the $E(F)$ MPL coherence check $S_0 + S_1 \stackrel{?}{=} S_2$. If no fault is injected, the above equation is always true and only then can the exponentiation result be released (after fault detection) by performing $S_0 + S_4$ calculation.

5.4 Security Analysis

The MPL algorithm due to its regularity in the number of O_i operations performed in its round, provides resistance against SSCAs (and more specifically PAs). Thus, simple PAs, the simplest form of horizontal SCAs, are not successful against MPL. The atomic block approach, that has been found to be vulnerable to advanced horizontal attacks, like Big Mac, HCA, HCCA attack [2, 3, 5, 6], is not applied in MPL (the algorithm uses no dummy data and is by design highly regular). However, some ASCA horizontal attacks can be successful even against MPL. This problem can be thwarted by the use of message/base point blinding (with a high bit length random element) and by avoiding the use of digit serial $\mathbb{Z}_n^*$ or F operations (mainly multiplications).

The adopted blinding technique of the proposed algorithm prevents vertical SSCAs (vertical SPAs) (like DA, RDA) since the connection between two consecutive messages/base point inputs is lost (they are blinded with different random numbers/points). However, message/base point blinding randomization, as indicated in [8], is not enough to provide protection against 2-TorA. So, it is imperative that the intermediate computation results are blinded with a different random element of $\mathbb{G}$ in every ME/SM round. This is achieved by exponentiating/scalar multiplying the random element B along with the message/base point without normalizing the random element to B at the end of each ME/SM round, as is done in similar blinding techniques (e.g., in the BRIP approach [19]).

The random element involvement in each of the proposed algorithm's round with out normalization (apart from the end of the algorithm) enhances message/base point blinding and makes the proposed approach highly resistant against ASCAs (and more specifically advanced PAs). DPA and CPA are not successful against the proposed message/base point blinding approach.

Regarding fault injection attacks, the proposed algorithms, as already mentioned, rely on the MPL round coherence check introduced at the end of a single ME or SM operation. This enhances the principle of fault infective computation introduced in [43]. However, a clever attacked could try to bypass the fault detection mechanism by introducing an additional fault after this function complementing an already injected fault during the main algorithmic process [31] (similar to the KQ attack). This two fault approach is not applicable in the proposed algorithm since the faulty result after fault detection remains blinded. Unblinding correctly this result will require a correct value (not faulty) to be used after fault detection. By bypassing the detection mechanism the attacker cannot discriminate if the ME/SM output is a blinded correct result or a faulty result. Thus, this result is useless for fault analysis.

5.5 Conclusion

In this book chapter, a common protection approach against SCA-PA and FA attacks is introduced both for RSA and ECC primitive operations of modular exponentiation and scalar multiplication, respectively. Our approach adopts and extends the MPL algorithm by introducing message/base point blinding, extension of the randomization operation per ME/SM round through a random element exponentiation/scalar multiplication in every round and infective computation along with a fault detection mechanism that releases the correct result only after passing the MPL coherency check. The proposed algorithmic solution constitutes a protection framework against a wide variety of SSCA and ASCA attacks (focusing on PAs) as well as FA attacks that introduce one or two faults and process them statically or statistically.

Acknowledgements This work is supported by EU COST action IC1204 "Trustworthy Manufacturing and Utilization of Secure Devices (TRUDEVICE)".

References

1. Fournaris A, Sklavos N. Public key cryptographic primitive design and protection against fault and power analysis attacks. In: DATE 2015 conference Workshop on trustworthy manufacturing and utilization of secure devices, 2015.
2. Bauer A, Jaulmes E, Prouff E, Wild J. Horizontal and vertical side-channel attacks against secure rsa implementations. In: Dawson E, editor. Topics in cryptology, CT-RSA 2013, ser. LNCS, vol. 7779. Berlin, Heidelberg: Springer; 2013. p. 1–17.
3. Bauer A, Jaulmes E, Prouff E, Wild J. Horizontal collision correlation attack on elliptic curves. In: Lange T, Lauter K, Lison KP Selected areas in cryptography—SAC 2013, ser. Lecture notes in computer science, vol. 8282. Berlin, Heidelberg: Springer; 2014. p. 553–70.
4. Koc CK. Cryptographic engineering. 1st ed. Incorporated: Springer Publishing Company; 2008.
5. Walter C. Sliding windows succumbs to big mac attack. In: Koc C, Naccache D, Paar C, editors. Cryptographic hardware and embedded systems CHES 2001, ser. Lecture notes in computer science, vol. 2162. Berlin, Heidelberg: Springer, 2001. p. 286–99.

6. Clavier C, Feix B, Gagnerot G, Roussellet M, Verneuil V. Horizontal correlation analysis on exponentiation. In: Soriano M, Qing S, Lpez J, editors. Information and communications security, ser. Lecture notes in computer science, vol. 6476. Berlin, Heidelberg: Springer; 2010. p. 46–61.
7. Fouque PA, Valette F. The doubling attack why upwards is better than downwards. In: Walter C, Koc C, Paar C, editors. Cryptographic hardware and embedded systems—CHES 2003, ser. Lecture notes in computer science, vol. 2779. Berlin/Heidelberg: Springer, p. 269–80.
8. Yen S, Ko L, Moon S, Ha J. Relative doubling attack against Montgomery Ladder. Inf Secur Cryptol. 2006;2005:117–28.
9. Yen SM, Lien WC, Moon SJ, Ha J. Power analysis by exploiting chosen message and internal collisions—vulnerability of checking mechanism for rsa-decryption. In: Dawson E, Vaudenay S, editors. Mycrypt, ser. Lecture notes in computer science, vol. 3715. Springer; 2005. p. 183–95.
10. Kocher P, Jaffe J, Jun B. Differential power analysis. In: Advances in cryptology proceedings of crypto 99. Springer; 1999, p. 388–97.
11. Amiel F, Feix B, Villegas K. Power analysis for secret recovering and reverse engineering of public key algorithms. In: Adams C, Miri A, Wiener M, editors. Selected areas in cryptography, ser. Lecture notes in computer science, vol. 4876. Berlin, Heidelberg, Springer; 2007. p. 110–25.
12. Bogdanov A, Kizhvatov I, Pyshkin A. Algebraic methods in side-channel collision attacks and practical collision detection. In: Chowdhury D, Rijmen V, Das A, editors. Progress in cryptology—INDOCRYPT 2008, ser. Lecture notes in computer science, vol. 5365. Berlin, Heidelberg: Springer; 2008. p. 251–65.
13. Moradi A. Statistical tools flavor side-channel collision attacks. In: Pointcheval D, Johansson T, editors. Advances in cryptology EUROCRYPT 2012, ser. Lecture notes in computer science, vol. 7237. Berlin, Heidelberg: Springer; 2012. p. 428–45.
14. Feix B, Roussellet M, Venelli A. Side-channel analysis on blinded regular scalar multiplications. In: Meier W, Mukhopadhyay D, editors. Progress in cryptology—INDOCRYPT 2014, ser. Lecture notes in computer science, vol. 8885. Springer International Publishing; 2014. p. 3–20.
15. Bauer A, Jaulmes I. Correlation analysis against protected sfm implementations of rsa. In: Paul G, Vaudenay S, editors. Progress in cryptology INDOCRYPT 2013, ser. Lecture notes in computer science, vol. 8250. Springer International Publishing; 2013. p. 98–115.
16. Joye M, Yen S-M. The montgomery powering ladder. In: CHES '02: revised papers from the 4th international workshop on cryptographic hardware and embedded systems. London, UK: Springer; 2003. p. 291–302.
17. Coron J-S. Resistance against differential power analysis for elliptic curve cryptosystems. In: Proceedings of the first international workshop on cryptographic hardware and embedded systems, ser. CHES '99. London, UK: Springer; 1999. p. 292–302.
18. Goubin L. A refined power-analysis attack on elliptic curve cryptosystems. In: Public key cryptographyPKC 2003, 2002. p. 199–211.
19. Mamiya H, Miyaji A, Morimoto H. Efficient countermeasures against RPA, DPA, and SPA. Crypt Hardware Embed Syst. 2004;3156:243–319.
20. Amiel F, Feix B. On the BRIP algorithms security for RSA. In: Information security theory and practices. Convergence and next generation networks: smart devices; May 2008.
21. Boneh D, DeMillo RA, Lipton R-J. On the importance of checking cryptographic protocols for faults (extended abstract). In: EUROCRYPT'97, 1997. p. 37–51.
22. Ciet M, Joye M. Elliptic curve cryptosystems in the presence of permanent and transient faults. Des Codes Crypt. 2005;36(1):33–43.
23. Fouque P-A, Lercier R, Réal D, Valette F. Fault attack on elliptic curve montgomery ladder implementation. In: 2008 5th workshop on fault diagnosis and tolerance in cryptography. IEEE; Aug. 2008. p. 92–8.
24. Shamir A. Method and apparatus for protecting public key schemes from timing and fault attacks. U.S. Patent 5,991,415, May 1999.

25. Sung-Ming Y, Kim S, Lim S, Moon S. RSA speedup with residue number system immune against hardware fault cryptanalysis, vol. 2288. In: Information security and cryptology ICISC 2001, 2002. p. 397–413.
26. Sung-Ming Y, Seungjoo K, Seongan L, Sang-Jae M. RSA speedup with chinese remainder theorem immune against hardware fault cryptanalysis. IEEE Trans Comput. 2003;52(4): 461–72.
27. Blömer J, Otto M, Seifert J. A new CRT-RSA algorithm secure against Bellcore attacks. In: Proceedings of the 10th ACM conference on computer and communications security. ACM, 2003. p. 311–20.
28. Wagner D. Cryptanalysis of a provably secure CRT-RSA algorithm. In: Proceedings of the 11th ACM conference on computer and communications security. ACM, 2004. p. 92–7.
29. Liu S, King B, Wang W. A CRT-RSA algorithm secure against hardware fault attacks. In: 2nd IEEE international symposium on dependable. Autonomic and secure computing, 2006. p. 51–60.
30. Qin B, Li M, Kong F. Further cryptanalysis of a provably secure CRT-RSA Algorithm. In: The 1st international symposium on data, privacy, and E-Commerce (ISDPE 2007). IEEE, Nov. 2007, p. 327–31.
31. Kim C, Quisquater J. Fault attacks for CRT based RSA: new attacks, new results, and new countermeasures. Smart cards, mobile and ubiquitous computing systems. Inf Secur Theory Pract. 2007;4462:215–28.
32. Fan J, Verbauwhede I. An updated survey on secure ECC implementations: attacks, countermeasures and cost. Crypt Secur From Theory Appl. 2012;6805:265–82.
33. Giraud C. An rsa implementation resistant to fault attacks and to simple power analysis. IEEE Trans Comput. 2006;55(9):1116–20.
34. Fumaroli G, Vigilant D. Blinded fault resistant exponentiation. In: Breveglieri L, Koren I, Naccache D, Seifert J-P, editors. FDTC, ser. LNCS, vol. 4236. Springer; 2006. p. 62–70.
35. Fournaris A, Koufopavlou O. Protecting crt rsa against fault and power side channel attacks. In: 2012 IEEE Computer Society Annual Symposium on, VLSI (ISVLSI, Aug. 2012. p. 159–64.
36. Amiel F, Villegas K, Feix B, Marcel L. Passive and active combined attacks: combining fault attacks and side channel analysis. In: Proceedings of the workshop on fault diagnosis and tolerance in cryptography, ser. FDTC '07. Washington, DC, USA: IEEE Computer Society; 2007. p. 92–102.
37. Schmidt JM, Tunstall M, Avanzi R, Kizhvatov I, Kasper T, Oswald D. Combined implementation attack resistant exponentiation. In: Abdalla M, Barreto P, editors. Progress in cryptology LATINCRYPT 2010, ser. Lecture notes in computer science, vol. 6212. Berlin, Heidelberg: Springer; 2010. p. 305–22.
38. Fournaris AP. Fault and simple power attack resistant rsa using montgomery modular multiplication. In: Proceedings of the IEEE international symposium on circuits and systems (ISCAS 2010). IEEE; 2010.
39. Kim CH, Quisquater JJ. How can we overcome both side channel analysis and fault attacks on RSA-CRT?. In: Workshop on fault diagnosis and tolerance in cryptography (FDTC 2007). IEEE; 2007. p. 21–9.
40. Boscher A, Handschuh H, Trichina E. Blinded fault resistant exponentiation revisited. In: Workshop on fault diagnosis and tolerance in cryptography (FDTC). IEEE; 2009. p. 3–9.
41. Danger JL, Guilley S, Bhasin S, Nassar M. Overview of dual rail with precharge logic styles to thwart implementation-level attacks on hardware cryptoprocessors. In: 2009 3rd international conference on, signals, circuits and systems (SCS). IEEE; 2009. p. 1–8.
42. Moradi A, Shalmani MTM, Salmasizadeh M. Dual-rail transition logic: a logic style for counteracting power analysis attacks. Comput Electr Eng. 2009;35(2):359–69.
43. Yen S-M, Kim S, Lim S, Moon S-J. Rsa speedup with chinese remainder theorem immune against hardware fault cryptanalysis. IEEE Trans Comput. 2003;52(4):461–72.

Chapter 6
Scan Design: Basics, Advancements, and Vulnerabilities

Samah Mohamed Saeed, Sk Subidh Ali and Ozgur Sinanoglu

6.1 Introduction

Security of Integrated Circuits (IC) is a major concern. Cryptochips, which apply encryption and decryption algorithms, are used in many applications such as cell phones, computers, avionics, smart cards, and medical applications to provide a secure environment. As any IC should be tested for defects, which are physical imperfection in the IC, to screened out defective chips, cryptochip can be hacked using the test features in the chip itself. Thus, cryptochip's test infrastructure can be turned into a backdoor to leak secret information of the chip.

Manufacturing test process targets ensuring a high level of quality and reliability of the chips with a minimum test cost. Providing a high test quality and low test cost is a major challenge in the test process. Test patterns are applied to detect faults, which represent defects at an abstracted functional level as a result of defects. To maximize the fault coverage, and, thus, the test quality, a large number of test patterns can be applied to detect as much defects as possible resulting in a large test data volume and, and thus, a long test time. The limited bandwidth as well as number of channels, which is used to transfer test data between the tester and the chip, can further prolong the test time. Although increasing the number of test channels can reduce the test time, it incurs higher tester cost. The end result is a high test cost. These interrelated challenges need to be tackled to ensure low-cost high-quality test.

S.M. Saeed (✉)
University of Washington, Tacoma, WA 98402, USA
e-mail: samahs@uw.edu; sms22@nyu.edu

S.S. Ali · O. Sinanoglu
New York University, 129188 Abu Dhabi, UAE
e-mail: sa11@nyu.edu; subidh.ali@nyu.edu

O. Sinanoglu
e-mail: os22@nyu.edu

N. Sklavos et al. (eds.), *Hardware Security and Trust*,
DOI 10.1007/978-3-319-44318-8_6

The semiconductor industry develops and adopts Design for Testability (DfT) techniques that modify the IC design, while maintaining its functionality. DfT techniques provide internal access to the chip, which includes controlling and observing the content of the storage elements to ensure a high quality. While DfT methods provide low-cost high-quality test, the IC is no longer secure against attackers that misuse the internal access to the IC to leak secret information from the chip. Throughout this chapter, we highlight the advanced DfT techniques for manufacturing test and shed light on the vulnerability of these techniques in security critical applications.

6.2 DfT

DfT [1] techniques enable comprehensive testing of the chip, enhancing the test quality. Unlike combinational circuits, in which a set of input combinations should be exercised to archive maximum fault coverage, sequential circuits, in addition, need to be traversed through all possible states. Thus, a sequence of test vectors may be required to detect any fault in a sequential circuit. However, having access to the primary inputs and outputs of the chip may be insufficient to cover all the states of the design, which can reduce the fault coverage, and, thus, the test quality. DfT modifies the design by adding hardware to enhance the test quality and minimize the test cost without affecting the functionality of the circuit itself. Testability, which represents the level of difficulty of testing internal signals in the design, is measured by controllability and observability of each signal line, where controllability measures the difficulty of setting a signal line to the required value, while observability measures the difficulty of propagating the logic value of a signal line to the output. DfT improves observability and controllability by providing access to the internal nodes of the design, which enhances the testability at the cost of limited hardware and performance overhead.

Many DfT techniques have been proposed to address the testing challenges. Structural DfT techniques, such as scan, partial scan, and boundary scan, are applicable to any circuit. Scan provides full access to the flip flops, turning them into scan cells, through the scan input/output pins so that the state of the design can be updated via shift-in operations. Partial scan provides full access to a selected subset of flip flops, providing a trade-off between area/performance overhead and testability. Boundary scan enables the test of the interconnect of logic using scan cells directly connected to the primary inputs and the outputs of the logical block. Next, we will describe in detail each one of these DfT approaches.

6.2.1 *Scan Design*

Scan design [1] is one of the most effective structured DfT solutions. It enables controlling and observing any internal state of the circuit. The scan design converts

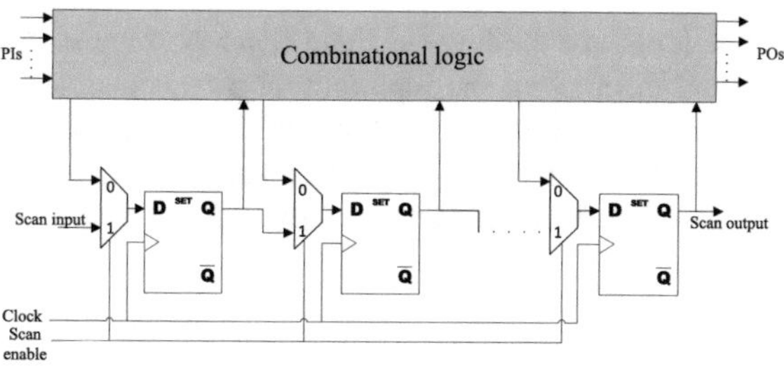

Fig. 6.1 Scan design

every flip flop to fully accessible scan cell by adding multiplexers to select either the output of the previous scan cell or the corresponding output of the combinational circuit to update each scan cell. All the scan cells, namely registers (flip flops), are linked together to form a chain, in which the first scan cell is driven by an input pin and the last scan cell drives an output pin. The scan design is illustrated in Fig. 6.1. If all the registers have the scan property, the design is considered as *full scan*. Otherwise, it is *partial scan*. While in the normal mode the chip performs its functional operations, the test mode in the scan design supports two different modes, which are the shift mode and the capture mode. Scan enable signal can be used to switch between these two modes. In the shift mode, the test stimulus is shifted into the scan chain through the scan input pin, while the test response is observed through the scan output pin one bit at a time. Shifting the test stimulus necessitates activating the shift mode until the whole pattern is shifted in. In the capture mode, the test stimulus already shifted into the scan cells is applied to the combinational logic circuit and then the test response is captured in the same scan cells. The captured test response can be observed, while shifting in a new stimulus pattern. As a result, sequential logic circuit can be treated as a combinational circuit, in which each flip flop can be treated as an input and an output at the same time. Therefore, the test quality is improved.

For larger designs with tremendous number of flip flops, shifting each test stimulus through a single scan chain results in a long test application time. A scan chain can be divided into many chains of shorter length as in Fig. 6.2, which can be accessed

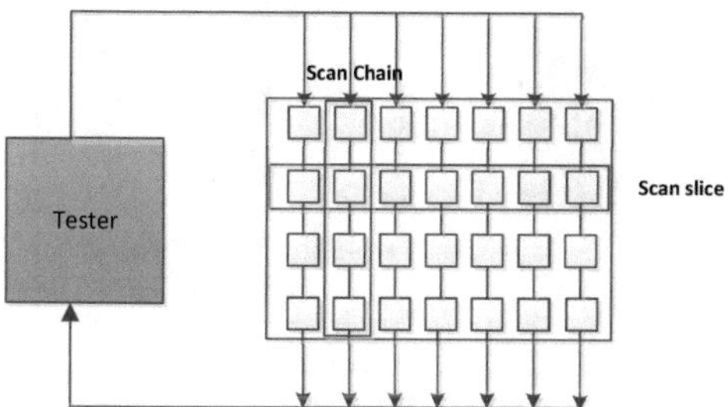

Fig. 6.2 Basic scan architecture: an example with 7 scan chains with a scan depth of 4

simultaneously. The length of the longest chain represents the scan depth. A group of scan cells of equal distance to the input/output pins is denoted as a scan slice. Increasing the number of chains reduces the scan depth, and, thus, the test application time at the cost of additional channels and pins, that are connected to the scan chains. Thus, there is a trade-off between the test time and the test cost.

6.2.1.1 Test Data Compression

Although the scan design enhances the testability, the test cost is dramatically increased for complex designs due to the long test time and the large tester memory requirement. To ensure high test quality, a large number of test stimulus and response patterns are stored. They occupy a large space on the external tester's storage. The storage capacity should be expanded to accommodate the larger number of patterns. The limitation of the bandwidth and the number of tester channels to transfer the test data between the tester and the chip increases the number of test cycles, and, thus, the overall test time. The test time can be reduced either through the reduction of the number of test patterns or the increase of the number of channels. However, the former one results in fault coverage loss, while the latter one is too costly to implement.

Test data compression [2–4] has been developed to address the problem of large test data volume and test time. Two components are added to the basic scan architecture, which are the stimulus decompressor and the response compactor. A stimulus decompressor expands a few number of tester channels into a much larger number of internal scan chains. A response compactor collects the responses from a large number of internal scan chains and feeds a small number of tester channels as illustrated in Fig. 6.3. Scan depth is reduced due to the increased number of internal scan chains while retaining the number of channels. As a result, the number of clock cycles for loading test stimuli and unloading the test responses is reduced, resulting

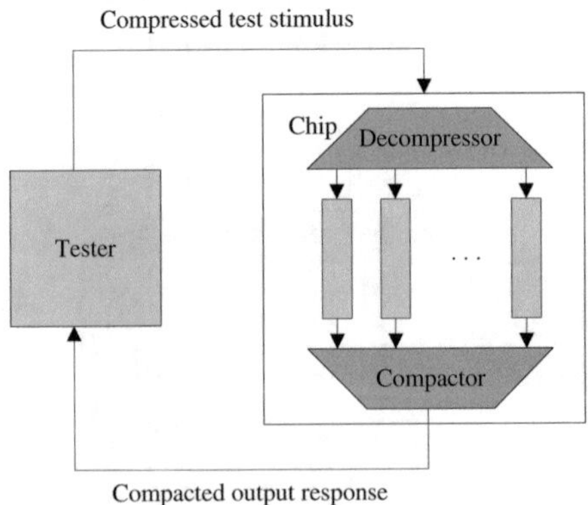

Fig. 6.3 Test data compression

in a reduction in the overall test time. Furthermore, since the size of each test vector is determined by the number of dedicated tester channels, the required tester storage is also reduced, resulting in a reduction in the overall test data volume. Therefore, test data compression reduces the test cost.

Test Stimulus Compression

Each test vector targets a specific set of faults. Only some bits of a test vector are utilized to activate and propagate the fault effects, while the remaining bits are left unspecified, referred to as *don't-care* bits. Test pattern generation tools can randomly specify these bits as 0's and 1's. A decompressor exploits the high density of don't-care bits in a stimulus (test pattern), compressing the test stimuli.

While adding a stimulus decompressor into the scan architecture reduces the test data volume, this scan architecture can degrade the test quality. The stimulus decompressor introduces correlation among the delivered bits to the chains, which depends on the decompressor structure. As a result, a stimulus decompressor maybe unable to deliver the desired test pattern; if a test pattern does not comply with the correlation induced by the decompressor, the test pattern is said to be unencodable. Faults that can only be detected by unencodable test patterns may remain untested in the presence of a stimulus decompressor. The internal structure of the decompressor determines the correlation, and, thus, delivery constraints.

A stimulus decompressor can be either sequential, such as Linear-Feedback Shift Register (LFSR), or combinational, such as fan-out and XOR-based decompressors [5]. An LFSR randomly generates the test pattern. Fan-out decompressors introduce correlation in the form of repeated bits within a slice fragment, whereas XOR-based decompressors introduce linear correlation among the bits delivered into scan cells. As shown in Fig. 6.4a, any 0–1 conflict within a slice fragment results in an unencodable pattern for the fan-out decompressor, as such a pattern fails to comply with the expected correlation. For XOR-based decompressors, the encodability of patterns is determined via solving a system of linear equations. Figure 6.4b provides an example of an unencodable pattern by highlighting the bits that result in unsolvable linear equations. In this figure, x's denote don't-care bits.

Typical test application procedures include a second phase, where unencodable patterns are applied serially by bypassing the decompressor [6]. As the second phase delivers no compression, every pattern applied in this phase degrades the overall compression level attained. Targeting an aggressive compression level, by increasing the number of internal scan chains, can reduce the test data volume per pattern in the first phase due to reduced scan depth. Yet, having to apply more patterns in the serial phase may offset the compression benefits of the first phase. A predictive analysis can help the designer in selecting the best possible configuration for a given compression technique at an early design stage, in order to find the balance between the test cost and the test quality.

Test Response Compaction

While a stimulus decompressor reduces the test data volume for the input stimuli, output responses can be similarly compressed by a response compactor. However, a response compactor may degrade observability. Some information is lost due to compaction, which can affect the observability and the fault coverage of the circuit. Some fault effects that were observed in the original circuit maybe masked due to output response compaction. The main underlying reasons are the *unknown values* and the *fault aliasing*. Unknown values can mask the fault effects captured in scan cells. Fault aliasing refers to the situation where multiple fault effects mask each other. An example of fault aliasing is illustrated in Fig. 6.5, where fault effects of $f1$ cancel each other upon getting compacted. Unknown values can be captured in the scan cells due to many reasons such as uninitialized memory and bus contentions. Unknown value, denoted by x, can mask the fault effects in the presence of response compactor. In Fig. 6.5, $f2$ is undetected, as its effect goes through the compactor along with an x. Although fault aliasing is a problem, the biggest concern is the unknown values due to their severe impact. Unknown values can be either static or dynamic [7, 8]. Static unknown values are discovered in the design time at the outputs of the un-modeled blocks (memory (RAM)) or bus contentions. Dynamic unknown values appear later after the design stage due to timing problems, the impact of operating parameters (voltage, temperature), and the defects caused during manufacturing.

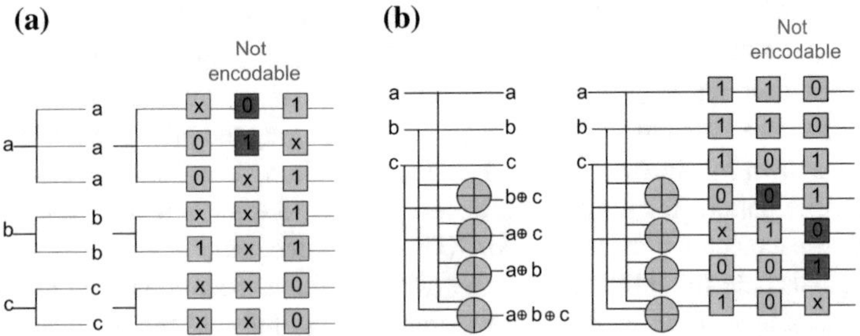

Fig. 6.4 Encodability problem. **a** Fan-out. **b** XOR decompressor

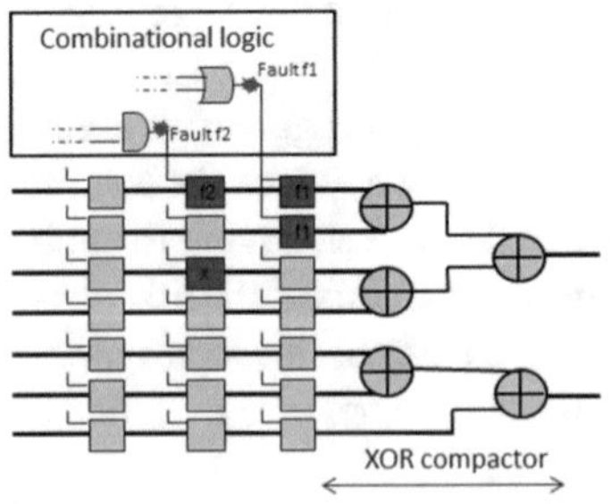

Fig. 6.5 The effect of XOR-compactor on fault coverage: fault cancellation and fault masking

Sequential compaction circuitries [9, 10], such as multiple input signature registers (MISRs), can be utilized for compressing the scan responses into a signature that is observed at the end of the test application process. The output vectors of the internal scan chains are compressed during different clock cycles to produce a signature that represents the output response of a certain pattern. A typical MISR consists of flip flops and XOR gates connected together into a register. MISR not only compresses a long scan-out sequence in the absence of unknown values, but also minimizes the aliasing impact on the fault coverage. However, one or more unknown values will corrupt the corresponding signature. Also, it is difficult to directly identify the location of the scan cell that captured a fault effect from the obtained signature of the MISR.

Combinational compaction solutions [11, 12], mostly XOR-based, are also utilized for response compaction. Every slice in the scan architecture is compacted independently. Unknown values may mask some of the captured bits in the same clock cycle, depending on the tolerance of a space compactor to unknown values per shift. However, the space compactor is susceptible to the occurrence of aliasing and offers reduced compaction levels than the time compactor.

Regardless of the compaction methods, unknown values can be handled in a variety of ways to achieve high fault coverage. Multiple XOR trees can be constructed that propagate the unknown values to the corresponding compressed response outputs, while observing scan cells that are connected to different compressed response outputs. Furthermore, DFT hardware can block unknown values before reaching the scan cells [13]. It is also possible to mask the unknown values before reaching the compactor [3, 14]. The response compactor can also be constructed to adapt to the varying density of unknown values in the response patterns. For XOR-based compactors for instance, the fan-out of scan chains to XOR trees within the compactor can be adjusted per pattern/region/slice to minimize the corruption impact of the unknown values in a cost-effective way [15, 16].

6.2.2 Boundary Scan

Boundary scan (also known as JTAG boundary scan) is a DFT technique, which is used to test interconnects, clusters of logic, and memory, while selectively overriding the functionality of each block of the logic circuits. The specification of the boundary scan was standardized as the IEEE standard 1149.1-1990 [17]. A boundary scan cell is connected to each input/output pin of a block. All the boundary scan cells in a block are linked serially to form a long shift register. The input of the shift register is called Test Data Input (TDI), while the output of the shift register is called Test Data Output (TDO). TDI and TDO represent the input and the output of a JTAG interface, respectively. A finite state machine, called Test Access Port (TAP) controller, controls all the possible boundary scan functions based on three signals, which are the external clock (TCK) signal, a Test Mode Select (TMS) signal, and an optional Test Reset (TRST) signal. The boundary scan architecture is illustrated in Fig. 6.6.

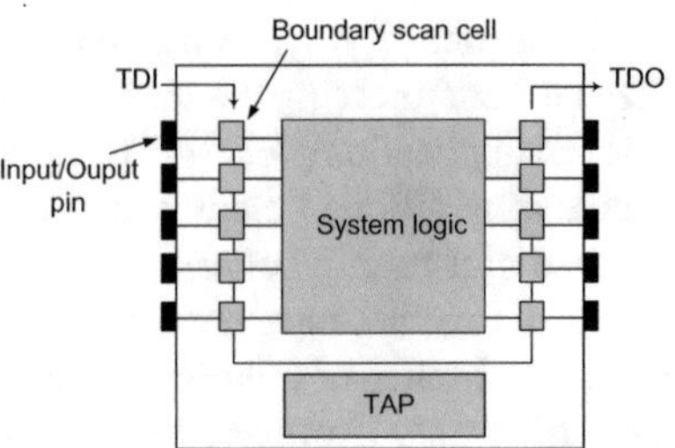

Fig. 6.6 Boundary scan architecture

The main advantage of the boundary scan architecture is the overall reduction in the number of input/output pins of the chip, as the external tester supports a limited number of tester channels. A serial two-pin interface helps access all the internal block inputs/outputs. In the normal mode, the input pins can directly feed the primary inputs of the logical block, while the primary output can be observed through the associated output pin. However, in the test mode, there is no longer a direct connection between the input/output pins of the chip and primary input/output of the logical block. The input/output pins of the chip are reused in the test mode as scan pins, which is a typical cost-effective implementation. During the shift mode, the test vectors can be serially shifted into the shift register through TDI, and response can be observed through the TDO. During the capture mode, the boundary scan cells drive the chip input pins and capture the chip output pins.

6.3 Scan-Based Side-Channel Attack

The scan design is an effective DfT technique that enhances the testability by providing full controllability and observability of the storage elements (flip flops) of the chip. However, the security may be compromised upon misuse of such capabilities. Scan design exposes the internal elements of the chip. Although some applications disable the scan chains after the manufacturing test by blowing fuses for example, other applications necessitate in-field testing to provide debug capabilities. For cryptochips, the scan design can be misused to leak the secret key of the chip. If the key register is part of the scan chain, the attacker can retrieve the key by simply shifting out the content of the scan chain. A good design practice is to exclude the key register from the scan chain. However, this alone does not guarantee a secure test environment. Scan-based side-channel attacks have been shown to leak secret information of the chip.

6.3.1 Attack Principle

Scan design can be exploited to circumvent the security of the chip. Some of the scan cells include secret information of the chip that executes encryption algorithms. The attacker targets the scan cells that store computation results of intermediate operations of the encryption algorithm. A scan-based side-channel attack utilizes the direct access to the primary inputs/outputs, and the scan-in/scan-out pins of the chip to recover the secret key; It uses the load and unload capabilities of the scan infrastructure. This attack applies differential analysis on different encryption algorithms such as Data encryption Standard (DES) [18] and Advanced Encryption Standard (AES) [19]. We will focus on the AES encryption algorithm throughout this chapter. However, our analysis can be extended to different encryption algorithms.

6.3.2 Advanced Encryption Standard (AES)

AES [20] is a well-known block cipher that supports block lengths of 128-bits and key lengths of 128, 192, and 256 bits. The AES algorithm consists of identical operations, i.e., rounds. The number of rounds depends on the key length; 10 rounds for 128-bit key, 12 rounds for 192-bit key and 14 rounds for 256-bit key. The AES encrypts the input, referred to as a *plaintext*, to the output, referred to as *ciphertext* after the desired number of rounds. The 128-bit input plaintext is represented as 4×4 matrix of input bytes, where each column is a separate word. Each round comprises the following four basic transformations, except for the last round, which omits `MixColumns`

- `SubBytes` (SB) is a nonlinear substitution operation. Each input byte to the SubBytes operation is replaced by another byte using one-byte substitution table, referred as S-box. This replacement is a one-to-one mapping.
- `ShiftRows` (SR) is the byte-wise permutation. The second, the third, and the fourth row of the matrix is cyclically shifted by one, two, and three positions to the left, respectively.
- `MixColumns` (MC) is a four-byte mixing operation. A linear transformation is applied to every column in the matrix, where each input byte in a column affects all the four bytes in the same column.
- `AddRoundKeys` (ARK) is XORing the state with the round key. Each output byte of the MixColumns operation is XORed with the corresponding key byte.

Figure 6.7 shows the structure of first round of AES, which contains an extra key XORing operation at the beginning. The intermediate results of every round is stored in the round registers.

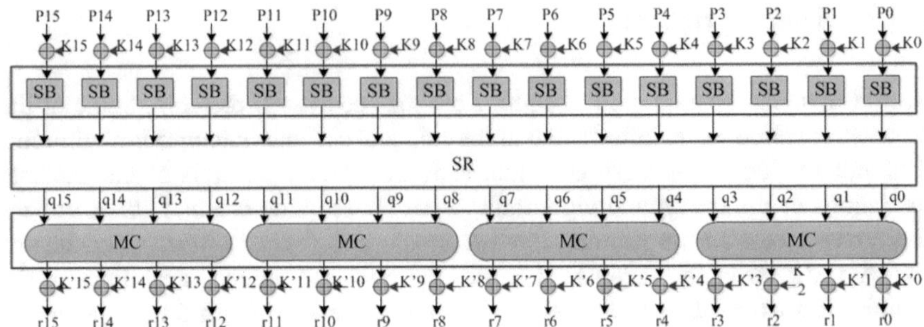

Fig. 6.7 First round of AES: p_i is the plaintext byte, k_i is the initial key byte, q_i is the *SR* output byte, k_i' is the round key byte, and r_i is the round output byte

6.3.2.1 Differential Properties of AES [21]

In AES S-box, for an input X and the input difference α, the output difference β is represented as

$$\beta = SB(X) \oplus SB(X \oplus \alpha) \tag{6.1}$$

For a given (α, β) pair, there could be no, two, or four solutions for X [22]. In the case of two solutions, they will be δ, and $\delta \oplus \alpha$, where δ is any nonzero solution for equation (6.1). In case of four solutions, they will be δ, $\delta \oplus \alpha$, 0 and α.

Lemma 1 *For a given input X and two nonzero differences α_i and α_j, the output differences β_i and β_j are*

$$\begin{aligned} \beta_i &= SB(X) \oplus SB(X \oplus \alpha_i) \\ \beta_j &= SB(X \oplus \alpha_j) \oplus SB(X \oplus \alpha_j \oplus \alpha_i) \end{aligned} \tag{6.2}$$

For any value X, β_i and β_j are distinct.

Proof We prove this by contradiction. Let as assume that there is a value x of X for which $\beta_i = \beta_j$. Let's define $y = x \oplus \alpha_j$. Then, we have two equations

$$\begin{aligned} \beta_i &= SB(x) \oplus SB(x \oplus \alpha_i) \\ \beta_j &= SB(y) \oplus SB(y \oplus \alpha_i), \end{aligned} \tag{6.3}$$

where $\beta_i = \beta_j$ implies that x and y are the two solutions of Eq. (6.1) where $\beta = \beta_i = \beta_j$ and $\alpha = \alpha_i$. Then either $y = x \oplus \alpha_i$, or x and y must be zero and α_i or vice versa. In either case, $\alpha_j = \alpha_i$ contradicting our assumption. Therefore, β_i and β_j must be distinct. □

6.3.3 Traditional Scan Attack

The traditional scan-based side-channel attack misuses the test infrastructure of the cipher [18, 19, 23–26]. As the round register is part of the scan chains, by switching from the normal mode to the test mode to observe the round register, the secret key can be recovered even if it is not included in the scan chain.

6.3.3.1 Attack Assumptions

The traditional scan attack works under the following assumptions:

- The details of the encryption algorithm running inside the cryptochip is known to the attacker.
- JTAG port, and, thus, the scan chains and the test capabilities, can be accessed by the attacker.
- The execution time for one round of the cipher is known to the attacker. Thus, the attacker can execute only one round operation and switch to the test mode.
- The registers for storing the round register key are not included in the scan chains.

6.3.3.2 Scan Attack on Basic Scan Architecture

With the chip in hand, the attacker can run the cipher in the functional mode with the desired plaintext for a few cycles, and then by switching to the test mode, he/she can shift out the content of the internal registers. These registers of the cryptochip hold the intermediate results of the cipher execution. Thus, the attacker can access the intermediate results of the cipher, and perform differential analysis on these results to get the secret key.

In traditional scan attacks on AES [19], the attacker first determines the scan chain architecture. The attacker identifies which bits belong to the round register as follows:

1. Apply a plaintext from the primary inputs in the functional mode, run the cipher for only one round in normal mode, then switch to test mode and shift out the contents of the scan chain. Let us call this output response f_1.
2. Repeat Step 1 for another plaintext with one-bit input difference, resulting in an output response f_2.
3. Compute the output difference of the previous two plaintexts ($f_1 XOR f_2$). The flip flops with a value of one correspond to the flip flops in the round register.
4. Repeat Steps 2 and 3 until all the flip flops of the round register are identified.

The previous steps of the attack identify each word of the round register; applying an input difference to a word affects only one word as per MixColumns properties.

The second step of the attack is to recover the round key. The attacker utilizes the basic differential property of AES, wherein among all possible S-box input pairs with

Table 6.1 The S-box input pair for each unique hamming distance

Unique HD	9	12	23	24
S-box input pair	(226, 227)	(242, 243)	(122, 123)	(130, 131)

one-bit difference in the least significant bit of a byte, only four pairs will produce an output difference of the round register with a unique hamming distance. Unique hamming distance refers to those hamming distances which correspond to a unique S-box input pair. These four unique hamming distances are 9, 12, 23, and 24.

Thus, to identify the round key byte, the attacker applies all possible 128 plaintext pairs with one-bit difference in the least significant bit of a byte, and observes the hamming distance in the captured round output. If a unique hamming distance is observed, he/she determines the corresponding unique S-box input pair. Table 6.1 shows the S-box input pair for each unique hamming distance, which is referred to as HD.

Thus, the corresponding two possible values of the key byte can be determined by just XORing the plaintext byte with each input of the S-box input pair. The same technique is applied across all the bytes to determine the final key. For each byte, attacker will obtain a pair of possible key byte values. Therefore, for all the sixteen bytes, the attacker will obtain 2^{16} possible 128-bit keys. In the worst case, the attacker has to apply $128 \cdot 16 = 2048$ plaintexts, while on average, 544 plaintexts are sufficient to retrieve the 128-bit AES key.

6.3.3.3 Scan Attack with Advanced DfT

Improved scan attacks have been proposed to adapt to the advanced DfT techniques such as partial scan [23], X-masking [24], and X-tolerant architecture [25, 26]. In the presence of test compression, the attacker may no longer able to observe the key-related flip flops (kffs) and compute the hamming distance of the output difference. Key-related flip flops are the flip flops of the round register that can be used to derive the secret key. Due to the presence of MixColumns operation, any byte of the input will affect only four bytes of the output. Thus, there are 32 kffs in AES. The effect of the response compactor on the scan attack depends on the distribution of the kffs in the scan architecture.

Let us consider a scan architecture with an XOR-compactor, in which each slice is compacted onto one channel. If each slice contains one kff as illustrated in Fig. 6.8a, the traditional scan attack can still reveal the secret key. When applying two plaintexts that differ in one byte, non-key-related flip flops will remain constant. Thus, the parity bit of each slice will be one if the value of the corresponding kff in the generated two responses is different and zero otherwise. On the other hand, when the kffs are distributed over at most 31 slices, at least one slice contains two kffs as shown in Fig. 6.8b. Thus, the hamming distance of the 32 kffs cannot be directly obtained by observing the response compactor output. Let us consider the worst case where all

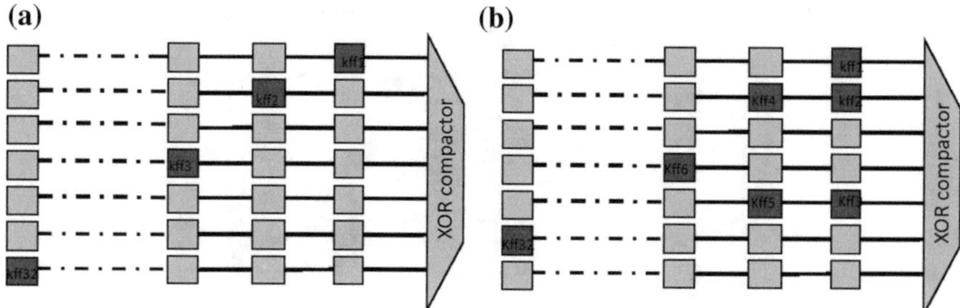

Fig. 6.8 Scan architecture with **a** one kff in each slice, **b** with more than one kff in some slices

the kffs are in the same slice. The modified scan attack is summarized in Algorithm 1 [25, 26].

Algorithm 1: Secret key recovery in the presence of XOR-compactor

- For each pair of plaintexts a_1 and a_2 that differ in the least significant bit ($a_1 = a_2 \oplus 1$)
  - Compute the output difference of the compacted responses of the two plaintexts (R_1 and R_2).
  - If $R_1 \oplus R_2 = 1$ (odd), consider the hamming distance 9 and 23. Otherwise, consider the hamming distance 12 and 24.
  - Compute the possible key byte using the corresponding S-box inputs and the plaintexts. (similar to the traditional attack).
  - Discard all the keys except the ones with the maximum occurrence k_1 (11 keys).
- Repeat the previous steps for each pair of plaintexts that differ in the second least significant bit ($a_1 = a_2 \oplus 2$) and compute k_2 (13 keys).
- Take the intersection of the two key sets to be the correct key ($k_1 \cap k_2$).

Thus, in the presence of an output response compactor, the modified scan attack is always able to derive the whole key with a complexity of 16 * (2^8 plaintexts) = 2^{12} = 4096 plaintexts and scan-out operations.

6.3.4 Test-Mode-Only Scan Attack

In order to retrieve the intermediate results of the cipher, the traditional scan attack has to rely on the condition that the intermediate results in the round register should be preserved upon a switch from the normal mode to the test mode. This condition can be easily eliminated by an automatic reset operation (*mode-reset countermeasure*) [27] upon a switch between the normal mode and the test mode. Therefore, all the existing scan attacks that rely on mode switching will fail in the presence of the reset operation.

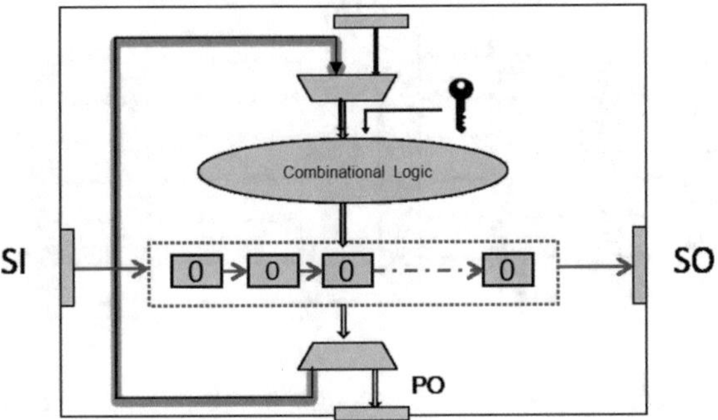

Fig. 6.9 Test-Mode-Only attack operations

A Test-Mode-Only scan attack [21, 28] has been proposed to circumvent the mode-reset countermeasure by staying in the test mode throughout the attack. In test mode, one can apply the plaintext or the intermediate input in the form of a test vector to the round operation of the target cipher and capture the corresponding response. This response is the round output corresponding to the applied test vector input. The test vectors are shifted in and the responses are shifted out through the scan input(*SI*) and the scan output (*SO*) pins respectively; load and unload capabilities of the scan infrastructure all within the test mode can be utilized for this purpose (Fig. 6.9).

6.3.4.1 Attack Assumptions

In addition to the assumptions of the traditional scan attack, the following are the assumptions in the Test-Mode-Only attack:

- In test mode, the user key is being used which is either hardcoded in the chip or stored in the memory.
- The global reset operation brings the chip to the first round by reseting the round counter.

6.3.4.2 Test-Mode-Only Attack on Basic Scan Architecture

An attacker has to mitigate the following challenges while developing a Test-Mode-Only attack.

1. Presence of boundary scan cells: In test mode, boundary scan cells drive the primary inputs. They block the direct access to the primary inputs through the chip input pins. Therefore, the attacker has only the *SI* pins to feed the cipher round.

2. Scan cell to round register flip flop mapping: The attacker does not know the mapping between the scan cells and the round register flip flops, as the physical placement tools decide how these flip flops are connected. As shown in [18], the flip flops are almost randomly connected by the physical placement tools. With an m-bit round register, there could be $m!$ possible mappings. For AES where the round register is 128 bits wide, the attacker has to try 128! possible mappings. We refer to the 128 scan cells that are associated with the round register as *key cells*.

The attacker utilizes the *SI* and *SO* pins to launch the attack. The attack is performed in four steps as the following:

Identifying the Key Cells

If an input difference in the AES round is applied, only the key cells will be affected, while the other scan cells preserve their content. Thus, the key cells can be distinguished from the other scan cells. Suppose we have a scan chain of n scan cells. We apply two test vectors V and V_i with one-bit difference at the i-th scan cell, and capture the responses. If we get a hamming distance greater than one in the difference corresponding to the output responses, then the scan cell i is a key cell. We vary i from 0 to $n - 1$, and determine all the 128 key cells.

Partitioning the Key Cells into AES Words

The second step of the attack relies on the fact that for AES, if an input difference is applied to a word, the bit-flips in the output difference will confine within only one word. It may be noted that in AES, the bytes in an input word and those in the corresponding output word are different. The output word bytes are those which get affected by the applied difference in the input word. In order to group the key cells into words, we apply two pairs of test vectors (V, V_i) and (V, V_j), where the one-bit difference is in the i-th and the j-th bit or key cell, respectively. If there is any common bit-flip in D_i and D_j, then i and j correspond to the same word of the round register. In that case all the bit-flips of D_i and D_j are in the same word. On average, 5 such input pairs are sufficient to determine all the bits of a word.

Partitioning the Key Cells into the AES Bytes

Based on the Differential property of AES S-box in Sect. 6.3.2.1, if we apply two different S-box input pairs A and B with the same input difference α, the output difference D_A and D_B should be different. Thus, we apply two pairs of test vectors, (V, V_i) and (V_j, V_{ij}) both with one-bit difference in the i-th bit, where V_{ij} is generated by flipping bits i and j in V. The output differences of the two pairs are different i

and j belong to the same byte. We fix i and vary all the 31 possible choices of j (in a word), and get the key cells corresponding to the byte of i. We repeat the procedure for different values of i, and group the key flip flips into the four bytes of the word.

The order of the key cells is unknown to the attacker in the corresponding byte. Therefore, the key recovery technique in [19] fails to retrieve the secret key.

Key Recovery Technique

The key recovery technique is based on a precomputed signature table. A signature table is created by applying eight one-bit differences corresponding to the eight bit positions of a S-box input byte. However, the differences are in the form (00000000, 00000001), i.e., an all-zero vector paired with a one-hot vector. There are eight such one-bit differences based on the location of the 1. The eight hamming distances at the one round output corresponding to the eight input differences are computed. These eight hamming distances are unique for a key byte value. A signature table for all the possible 256 values of the key byte is created. For each key byte value, each one of the eight one-bit differences corresponding to eight key cells of a byte is applied and the hamming distance of the output is computed. We compare the output hamming distance with each row of the signature table. The correct key is the key associated with the matching row. This attack requires only 9 test vectors to recover a key byte.

6.3.4.3 Test-Mode-Only Attack with Advanced DfT

In this section, we highlight the test mode-only-attack in the presence of the stimulus decompressor [29] as an example of a Test-Mode-Only attack in the presence of advanced DfT.

The stimulus decompressor imposes additional deliverability challenge on the Test-Mode-Only scan attack. Unlike the traditional scan attack in which the attacker applies the plaintext through the primary inputs in the normal mode, in the Test-Mode-Only attack, *SI* is the only input pin to load data, necessitating the data be loaded through the decompressor. The stimulus decompressor expands compressed data into bits delivered into scan cells, complicating the identification of the mapping between the flip flops and the corresponding inputs to the AES round. Based on the Test-Mode-Only attack on basic scan architecture, the attacker needs to apply independent bit-flip to each key cell. However, the stimulus decompressor could unintentionally flip other scan cells in the same slice which may lead to an erroneous result in the attack. This is analogous to the test pattern encodablity problem explained in Sect. Test Stimulus Compression, in which key cells in test patterns can be treated as care bits, while non-key cells can be considered as don't cares. Thus, the decompressor with higher test pattern encodability leads to a scan attack that is more likely to be successful.

For the attack to be successful, the key cells of a word should be distributed such that

- One key byte of the word should be *fully controllable*, which can take any of the 2^8 possible values.
- Each one of the remaining three bytes, can take at least two values, while the other bytes remain constant.

The following subsections describe the attack procedure.

Determining the Mapping with Multiple Correlated Key Cells

In AES, if one byte difference a is applied at the input of MixColumns operation, the output difference spreads to four bytes, where the output bytes show difference values of $2a$, a, a, and $3a$. The byte where the difference is applied will always receive an output difference of $2a$ (we refer to this byte as the "$2a$ byte"). Table 6.2 shows the four byte hamming distances in a sorted order; every column in this table corresponds to one of 256 values of a.

The proposed attack relies on certain properties in this table; this means that the scan attack should have the capability to apply all possible values of a, which requires the identification of the controllable and the corresponding key value through the existing Test-Mode-Only attack [21]. To be able to apply all possible values of a, we need to apply all possible patterns V from the scan cells corresponding to the $2a$ byte, as the S-box output difference is given by $a = SB(k) \oplus SB(k \oplus V)$, where k is the key byte value.

Next, we target the other three bytes (a, a and $3a$). In this attack, $2a$ byte captures responses of the byte with one-bit difference; these are the first seven columns of Table 6.2, for which Table 6.3 provides the actual difference values for the four bytes ($2a$, a, a, $3a$). The reason why we focus only on these seven columns of Table 6.2 is (1) the bytes $2a$, a and a all show a single bit-flip, (2) it is easier to distinguish byte $3a$ from other bytes by observing the bit-flip repetitions (e.g., the repetition of the bit-flip in the fourth bit from the left in the top two rows of Table 6.3 hint that this bit must belong to byte $3a$). This way, all the bits of the $3a$ byte except for the leftmost and the rightmost bits can be identified.

To identify the leftmost and the rightmost bits of the $3a$ byte, we choose two values of a such that only one bit (leftmost or rightmost bit of the $3a$ byte) repeats. To identify the leftmost bit of the $3a$ byte, for instance, we first apply the $2a$ value of 10000000 (fourth row of Table 6.3), followed up by the $2a$ value of 00010111. This

Table 6.2 Sorted Hamming distances corresponding to four bytes of a word, when one byte difference a is applied at the S-box output

$2a$	1	1	1	1	1	1	1	1	2	...	7	7	8
a	1	1	1	1	1	1	1	1	2	...	6	7	5
a	1	1	1	1	1	1	1	1	2	...	6	7	5
$3a$	2	2	2	2	2	2	2	4	2	...	3	2	3

Table 6.3 Actual difference values for the four bytes corresponding to the first seven columns in Table 6.2

$2a$	a	a	$3a$
00100000	00010000	00010000	00110000
00010000	00001000	00001000	00011000
00000100	00000010	00000010	00000110
10000000	01000000	01000000	11000000
00000010	00000001	00000001	00000011
01000000	00100000	00100000	01100000
00001000	00000100	00000100	00001100

would result in 11000000 followed by 10010001 in the $3a$ byte, creating a repetition that can be used to identify the leftmost bit of the byte. The rightmost bit of the $3a$ byte can be similarly identified.

Next, we target the two remaining bytes, a and a, and try to distinguish between them. As shown in Table 6.3, the two bytes show identical behavior. As the $2a$ and $3a$ bytes have already been identified, any remaining bit-flips are known to belong to one of these two bytes; the position of each of the remaining key bits can also be discovered, but which one of the a byte they belong to remains ambiguous. For instance, for $2a$ = 00100000 (first row), any observed bit-flip in the remaining two bytes is known to correspond to the fourth bit position from the left; however, there will be two such bit-flips, and which one of the two a bytes each bit belongs to will not be known. This way, although the position of all the bits in the a bytes can be identified, the bits of the two a bytes cannot be differentiated.

Recovering the Key

The key value for the $2a$ byte was already identified in the first step. Next, the byte that was identified as the $3a$ is determined by applying any nonzero difference from the corresponding scan cells of the byte and observe the output difference in the same byte as in Sect. 6.3.2.1.

For the remaining two bytes identified as the a bytes, the challenge is that the first step was not able to accurately classify the 16 bits into these two bytes. Nonetheless, by applying any nonzero difference as long as the difference is contained in one of the bytes (second condition of the attack), observing the output difference in the already identified bytes ($2a$ and $3a$ bytes) will determine: (1) which of the a bytes the difference was applied from based on the relationship between the content of the already identified bytes (they can be either identical, or one can be thrice the other), and (2) the actual value of the output difference. From the input and the output differences, and the knowledge of which byte the difference was applied from, the key value can be recovered for this byte. The same operation can be repeated for the other a byte to recover its key.

For the first step of the attack 32 bits are required to identify the $2a$ byte. To identify the mapping of the remaining three bytes 2^8 test vectors are applied. Therefore, the time complexity of the first step of the attack is 2^8, and $256 + 32 = 288$ test vectors are required. To recover the key byte of the rest of the three key bytes, 3 pairs of test vectors are required. Altogether, we need $(288 + 6) \cdot 4 = 1176$ test vectors. In the second part of the attack we know the circular order of the four bytes. Therefore, we have four possible permutations of the four bytes. The search space of the key word is $2^3 \cdot 4 = 2^5$. The search space of the entire key is given by $(2^5)^4 \cdot 4! = 2^{24.5}$, which is also the time complexity of the attack.

6.4 Summary

The interdependence between testability and security is receiving a lot of attention. While the manufacturing test necessitates deep access into the IC to enhance its testability, this can inadvertently threaten the security of the IC in security critical applications. On the other hand, although black box testing ensures security, it fails to deliver a high-quality test.

We describe various DfT techniques that address the test challenges. These techniques reduce the tester-induced costs. Then, we show the security vulnerability of scan-based DfT techniques. We review a few scan attacks that target the basic scan architecture as well as the compression-based scan architecture. We analyze the limitations of the proposed attacks, hinting at ways to design testable yet secure DfT.

References

1. Bushnell M, Agrawal V. Essentials of electronic testing for digital. Memory and mixed-signal VLSI circuits. Springer; 2005.
2. Rajski J, Tyszer J, Kassab M, Mukherjee N, Thompson R, Tsai KH, et al. Embedded deterministic test for low cost manufacturing test. In: Proceedings of IEEE international test conference, 2002. p. 301–10.
3. Barnhart C, Brunkhorst V, Distler F, Farnsworth O, Keller B, Koenemann B. OPMISR: the foundation for compressed ATPG vectors. In: Proceedings of IEEE international test conference, 2001. p. 748–57.
4. Samaranayake S, Gizdarski E, Sitchinava N, Neuveux F, Kapur R, Williams TW. A reconfigurable shared scan-in architecture. In: Proceedings of IEEE VLSI test symposium, 2003. p. 9–14.
5. Touba NA. Survey of test vector compression techniques. IEEE Des Test Comput. 2006;23(4):294–303.
6. Pandey AR, Patel JH. An incremental algorithm for test generation in illinois scan architecture based designs. In: Proceedings of design, automation and test in Europe conference and exhibition, 2002. p. 368–75.
7. Breuer MA. A note on three-valued logic simulation. IEEE Trans Comput. 1972;21(4):399–402.

8. IEEE standard hardware description language based on the verilog(r) hardware description language. IEEE Std 1364–1995, 1996. p. 1–688.
9. Savir J. Reducing the misr size. IEEE Trans Comput. 1996;45(8):930–8.
10. Rajski W, Rajski J. Modular compactor of test responses. In: Proceedings of IEEE VLSI test symposium, 2006. p. 10.
11. Pouya B, Touba NA. Synthesis of zero-aliasing elementary-tree space compactors. In: Proceedings of IEEE VLSI test symposium, 1998. p. 70–7.
12. Mitra S, Kim KS. X-compact: an efficient response compaction technique for test cost reduction. In: Proceedings of IEEE international test conference, 2002. p. 311–20.
13. Wohl P, Waicukauski JA, Ramnath S. Fully x-tolerant combinational scan compression. In: Proceedings IEEE international test conference, Oct 2007. p. 1–10.
14. Chickermane V, Foutz B, Keller B. Channel masking synthesis for efficient on-chip test compression. In: Proceedings of IEEE international test conference, 2004. p. 452–61.
15. Saeed SM, Sinanoglu O. Multi-modal response compaction adaptive to x-density variation. IET Comput Dig Techniq. 2012;6(2):69–77.
16. Saeed SM, Sinanoglu O. Xor-based response compactor adaptive to x-density variation. In: Proceedings of IEEE Asian test symposium, 2010. p. 212–17.
17. IEEE standard test access port and boundary scan architecture. IEEE Std 1149.1-2001, July 2001. p. 1–212.
18. Yang B, Wu K, Karri R. Scan based side channel attack on dedicated hardware implementations of data encryption standard. In: Proceedings of IEEE international test conference, 2004. p. 339–44.
19. Yang B, Wu K, Karri R. Secure scan: a design-for-test architecture for crypto chips. In: Joyner Jr. WH, Martin G, Kahng AB, editors. ACM/IEEE design automation conference; 2005. p. 135–40.
20. Daemen J, Rijmen V. The design of Rijndael. New York: Springer Inc.; 2002.
21. Ali SS, Sinanoglu O, Saeed SM, Karri R. New scan-based attack using only the test mode. In: Proceeding of IEEE VLSI-SoC, 2013. p. 234–39.
22. Nyberg K. Generalized feistel networks. In: Kim K, Mat-Sumoto T, editors. ASIACRYPT, volume 1163 of lecture notes in computer science. Springer; 1996. p. 91–104.
23. Kapur R. Security vs. test quality: are they mutually exclusive? In: Proceeding IEEE test conference, 2004. p. 1414.
24. DaRolt J, Di Natale G, Flottes ML, Rouzeyre B. Are advanced DfT structures sufficient for preventing scan-attacks? In: Proceedings of IEEE VLSI test symposium, 2012. p. 246–51.
25. Ege B, Das A, Ghosh S, Verbauwhede I. Differential scan attack on AES with X-tolerant and X-masked test response compactor. In: IEEE DSD, 2012. p. 545–52
26. DaRolt J, Di Natale G, Flottes ML, Rouzeyre B. Scan attacks and countermeasures in presence of scan response compactors. In: Proceeding of European test symposium, 2011. p. 19–24.
27. Hely D, Bancel F, Flottes ML, Rouzeyre B. Test control for secure scan designs. In: Proceedings of IEEE European symposium on test, 2005. p. 190–5.
28. Ali SS, Saeed SM, Sinanoglu O, Karri R. Scan attack in presence of mode- reset countermeasure. In: Proceeding of IEEE international on-line testing symposium, 2013. p. 230–1.
29. Saeed SM, Ali SS, Sinanoglu O, Karri R. Test-mode-only scan attack and countermeasure for contemporary scan architectures. In: Proceedings of IEEE international test conference, 2014. p. 1–8.

Chapter 7
Manufacturing Testing and Security Countermeasures

Giorgio Di Natale, Marie-Lise Flottes, Bruno Rouzeyre and Paul-Henri Pugliesi-Conti

7.1 Introduction

As described in the previous chapter, manufacturing test is the only process able to ensure quality and reliability of manufactured integrated circuits. The fastest and best cost-effective solution for digital testing is based on the use of scan chains. Unfortunately, this solution might allow a malicious user to exploit this test infrastructure and retrieve secret information stored within the integrated circuit (see previous chapter). The antagonism between scan-based Design-for-Testability (DfT) and security comes from their competing goals: improving controllability and observability of internal states for increased testability, and preventing control or observation of these internal states for increased security.

In this chapter, we describe solutions from the literature to counteract possible attacks targeting malicious usage of scan chains and, more generally, test infrastructures. Moreover, we present industrial practices and potential downsides when implementing secure test infrastructures. Because increased security should not be achieved at the detriment of product quality, we discuss potential testability loss when secure-test approaches impacts the test procedure and expected feedback compared to common practices.

Section 7.2 classifies countermeasures to test-based attacks according to the strategy, i.e., using a secure control/usage of the embedded test infrastructure, deleting the access to the test infrastructure by shifting test resources to the device under test (DUT), or deleting the test infrastructure itself by changing the test approach from structural to functional testing. According to this classification, the following sections provide deeper analysis and implementation details of major countermeasures

G. Di Natale (✉) · M.-L. Flottes · B. Rouzeyre
LIRMM (Université Montpellier II/CNRS UMR 5506), Montpellier, France
e-mail: giorgio.dinatale@lirmm.fr

P.-H. Pugliesi-Conti
NXP Semiconductors Caen, Caen, France

N. Sklavos et al. (eds.), *Hardware Security and Trust*,
DOI 10.1007/978-3-319-44318-8_7

proposed in the literature. In particular, Sect. 7.3 analyzes built-in test solutions where the totality or only a part of the test resources as test pattern generation and/or test response analysis are embedded in the DUT instead of being outsourced from/to an external Automatic test equipment. Section 7.4 reports the major contributions in the field of secure test access mechanisms in order to provide a comprehensive view of all solutions. Section 7.5 discusses security and applicability of the DfT of some selected solutions from an industrial point of view. Eventually, Sect. 7.6 concludes this chapter.

7.2 Countermeasures to Scan-Based Attacks

Several countermeasures to scan-based attacks have been proposed in the literature and in the industry. We classify them in the following categories:

1. **Functional Test**. In this solution, no test infrastructure are embedded in the circuit and its test is guarantee by applying functional test pattern through the primary inputs of the circuit. Functional testing approaches are based on a functional model of the system [1]; they attempt to reduce the complexity of the test generation problem of structural approaches by using higher level of abstraction. However, modeling complex systems at high-level remains difficult. (Pseudo-)Exhaustive testing, that assumes that any permanent fault is possible (implicit fault model) suits well with regular structures but not with arbitrary circuits, which requires either long exhaustive and pseudo-exhaustive sequences or pre-partitioning. Explicit functional fault models are likely to produce prohibitively large set of faults. Implicit functional fault models have been however successfully used in testing RAMs or microprocessors where test patterns can be developed as a sequence of instructions. Nevertheless, implicit fault models cannot be exploited for the generation of test patterns for generic circuits. Lastly, functional test effectiveness is difficult to evaluate.
2. **Physically disconnecting the test infrastructures**. A common technique (especially adopted by smart card providers) is to disable or disconnect the test circuitry after manufacturing test by blowing fuses located at the ends of the scan chains [2], as shown in Fig. 7.1.
   The main advantage is that the circuit can be fully tested at manufacturing time, while the test mode cannot be accessed anymore after it. However, in-field maintenance and debug are compromise afterward. Moreover, invasive attacks based on microprobing can re-build the test connection.
3. **Built-In Self-Test**. Within this solution, the circuit is self-tested thanks to an extra test pattern generator (TPG) and an output response analyzer (ORA) embedded with the DUT. Scan chains still exist in the design for sequential circuits, but Scan-In and Scan-Out I/Os are not accessible from the DUT interface thus preventing control and observation of the DUT internal states through the test infrastructure. This solution can achieve high level of testability with no visible scan chain and

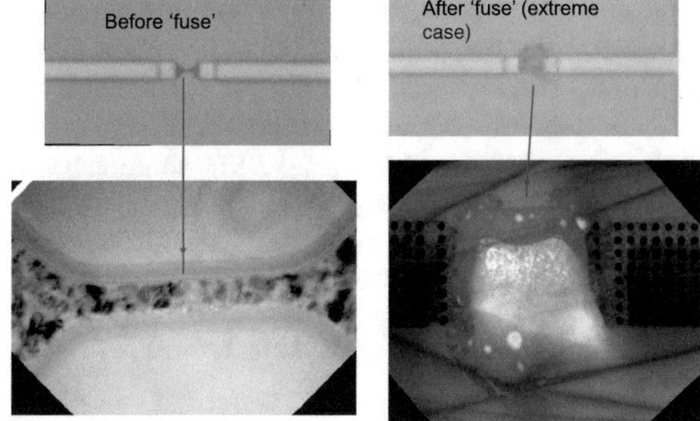

Fig. 7.1 Poly Fuse: Detailed inspection: CMOS090 TEM *top down view*

negligible area overhead but it must be set and evaluated on a case-by-case basis. On the other hand, BIST does not allow diagnosing possible fault location within the circuit. More details about this solution are provided in Sect. 3.1.

4. **On-Chip Test Comparison**. This approach is based on the concept of withholding information. Scan-based attacks are based on the observation of the scan chain content, which is shifted-out after executing a part of the encryption algorithm, i.e., when the scan chains contains information that can be exploited to retrieve the secret key. The idea is thus to compare test responses within the chip, so that no secret information can be observed through scan-out operations. Both input vectors and expected responses are scanned into the circuit and the comparison between expected and actual responses is done vector-wise, so that it does not provide information on the value of each individual scan bit for security purposes. More details about this solution are provided in Sect. 3.2.
5. **Secure Test Access Mechanism**. Many related approaches have been explored. They consider either the use of standard test interface with Scan-In, Scan-Out, and Scan-Enable signals or with nonstandard scan designs. In the first case, the test interface is enabled iff an initialization or authentication mechanism is performed earlier. After the execution of a secure protocol, the access to the scan chain is granted. In the second case, nonstandard test interfaces are considered, from which it is not possible to extract useful information and recover the encryption. More details about these solutions are provided in Sect. 7.4.

7.3 Built-In Self-Test

One approach for providing test solutions at different stages of an IC life cycle consists in including built-in self-test (BIST) resources into the DUT. Classically, storage elements are organized into scan chains and additional hardware is used for feeding the scan chains with pseudorandom test data, and sinking the test responses before analysis of the compressed signature [3]. Therefore, BIST does not provide controllability and observability of the internal storage elements from the circuit interface. However, BIST must be implemented at low cost and its efficiency must be demonstrated in terms of fault coverage and test duration.

In the next subsections, we propose dedicated built-in self-test solutions for cryptographic cores, as well as partial-BIST solutions, where the test generation is performed by the external tester while the response analysis is performed on-chip.

7.3.1 BISTed Cryptographic Cores

Random pattern testability of crypto-cores has been discussed in [4]. Authors show how random data and possible errors can be easily propagated through typical operations involved in encryption algorithms.

Security provided by block cipher algorithms such as DES and AES relies on two main properties named *diffusion* and *confusion* [5, 6]. Confusion refers to making the relationship between the key and the ciphertext as complex and involved as possible. Diffusion refers to the property that redundancy in the statistics of the plaintext is dissipated in the statistics of the ciphertext. For diffusion to occur, a change in a single bit of the plaintext should result in changing the value of many ciphertext bits. These properties are supported by the Feistel network [7] for the DES and by the substitution—permutation network for the AES. AES and DES also have two common characteristics. First, they are iterative algorithms. DES is composed of 16 rounds while AES is made of 10 rounds. All rounds are (quasi) identical, i.e., the result of a round is used as the input of the next round. Second, since encryption/decryption are bijective operations for a given key, each round is a bijective operation too (on a set of 2^{64} elements for DES on a set of 2^{128} elements for AES).

The diffusion property is a very interesting feature with regard to the test of their hardware implementation. It implies that every input bit of the round module influences many output bits, i.e., every input line of a round is in the logic cone of many output bits. In other words, an error caused by a fault in the body of the round is very likely to propagate to the output. Thus, the circuit is very observable. Moreover, since rounds are bijective, the input logic cone of every output contains many inputs. In other words, each fault is highly controllable. Therefore, these circuits are highly testable by nature whatever the implementations.

Example of BIST implementations are provided in [8, 9]. Figure 7.2 (except for yellow area) presents a generic implementation of either AES [10] or DES [11] symmetric cryptographic algorithms. The hardware implementation is mainly composed of a key-generation module and a Round module. In mission mode, after an initial operation (XOR between Key and Plaintext for AES, and permutation of the plaintext for DES), the plaintext block is looped around the Round module several times (10 for AES, 16 for DES) before the final cipher is loaded into the output register, possibly after a final operation like the final permutation in DES. The yellow area in Fig. 7.2 depicts the required modifications to support the built-in self-test of the module itself.

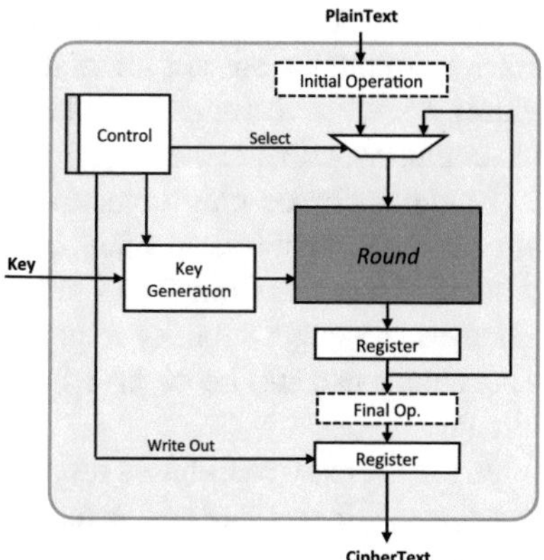

Fig. 7.2 Implementation of the symmetric cryptographic algorithm

During the BIST mode, an initial message M_1 is encrypted into $M_2 = \text{Round}(M_1)$ and the process is repeated n times ($M_{i+1} = \text{Round}(M_i)$, $i \in \{1 \ldots n\}$). Finally, the output data M_{n+1} is stored into the output register for comparison with the expected golden value. Concerning round key generation, either the keys are precomputed and stored in the circuit or the key generation module calculates the sequence of keys. For the latter case, AES is modified in such a way that the tenth round key is used as the primary key for the next round key generation. In this way, during self-test, the key-generation module receives as many different stimuli as rounds. For DES, this is not necessary because the key-generation module does not contain any logic. The round keys are simply formed of subsets of bits of the initial key.

It has been shown that for DES, with several keys and initial input messages, after 25 encryptions (i.e., 400 clock cycles), the whole circuit (round module and control module) has always been fully tested [8]. In the same way, for AES the experiments have been repeated with different plaintexts and secret keys as starting points, obtaining test sequences ranging from 2100 to 2500 patterns [9].

Following the same principle, in [12] the authors propose a solution for the BIST of public-key cryptocores. As the modular multiplication is at the heart of many public-key algorithms, they considered the Elliptic Curve Cryptography (ECC) as the appropriate choice for the public-key cryptosystem. The key idea is to configure the multiplier such that it concurrently acts as both a test pattern generator and signature analyzer. As in the previous solution, the outputs are fed back to the inputs providing the test patterns. Concurrently, the multiplier compacts its outputs to the final signature. Experimental results showed that very high fault coverage can be obtained with a very limited number of clock cycles.

7.3.2 *Built-In Test Comparison*

Secure on-chip test comparison has been proposed as a solution to eliminate the need of disconnecting the scan chains [13–15]. This approach is based on the concept of withholding information. This approach is not, strictly speaking, a BIST solution since test patterns are not internally generated. The test procedure consists in providing both test vectors and expected test responses to the DUT and in performing the comparison inside the chip.

Methods for on-chip comparison of actual and expected test responses have already been explored in other contexts [16, 17], mainly to reduce the test data volume to transfer from DUTs to test equipment. However, none of these solutions achieves the target security requirements since individual bit values stored in the scan chains can still be observed or deducted from observed data thanks to the test circuitry.

In the standard scan-based test mechanism, Flip-Flops (FFs) are replaced by scan flip-flops (SFF) and are connected so that they behave as a shift register in test mode. The output of one SFF is connected to the input of next SFF. The input of the first FF in the chain is directly connected to an input pin (Scan-In) while the output of the last FF is directly connected to an output pin (Scan-Out). An additional signal (Scan-Enable) selects whether SFFs have to behave normally or as a shift register. The test procedure is composed of three steps: first, test patterns are shifted-in via the scan chain (i.e., by keeping Scan-Enable = 1) for #SFF clock cycles (where #SFF is the number of SFFs in the chain). Second, one or two functional clocks (i.e., Scan-Enable = 0) are applied to capture the circuit's response. Usually, one clock cycle is used for static faults, while 2 (or even more) clock cycles are used for dynamic faults. Finally, the content of SFFs is shifted out for #SFF clock cycles (again, with Scan-Enable = 1) to allow the ATE to compare the obtained values with respect to the expected ones.

The principle of the approach proposed in this solution is to compare the actual responses with the expected ones within the chip boundaries instead of scanning-out the actual responses and comparing it within the ATE. In order to guarantee that secure data cannot leak outside the chip, the output of the comparison is not bitwise delivered to the ATE, but only after applying and comparing the whole test vector (i.e., after comparing the value of each SFF). Therefore, a potential attacker can no longer observe the FFs content but simply pass/fail information for the whole test vector.

The general scheme of the proposed secure comparator is shown in Fig. 7.3. Instead of directly shifting DUT's responses (Scan-Out) out of the chip, the ATE also provides the expected responses using the Sexp pin and the actual test response is on-chip compared with the expected one. After having compared all #SFF bits captured in the scan chain, the signal TestRes is asserted if the whole test response matches the one with expected values. Indeed, if the result of the comparison was accessible at each clock cycle instead of each test vector, an attacker could easily observe the scan chain content by shifting in "000…000" on the Sexp pin. Each bit-comparison would then validate that either the actual bit was '0' when TestRes = 1 and vice-versa.

The Secure Comparator is composed of three parts: the *Sticky Comparator* responsible for the comparison between the bitstream coming from the scan chains and the

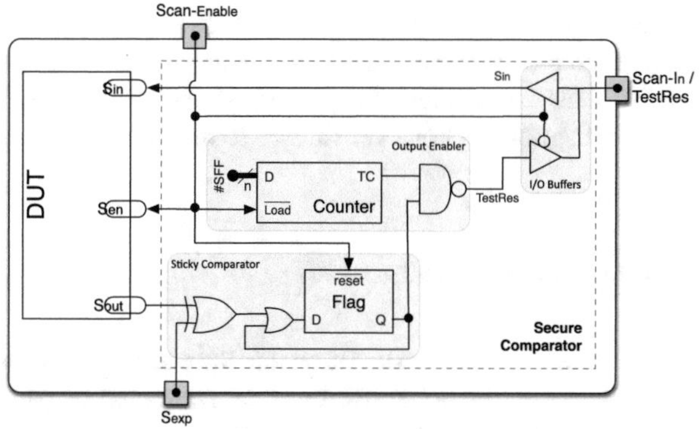

Fig. 7.3 Secure comparator

expected values; the *Output Enabler* triggering the final comparison result; the *I/O Buffers* allow keeping the test pin count as in a classic scan-based approach.

The sticky comparator performs a bitwise serial comparison between the bitstream coming from S_{out} and the one from S_{exp}. A FF (Flag in the figure) is initially reset and then it rises to '1' whenever one comparison fails. The reset of the flag is performed when the scan operation is not enabled (i.e., S_{en} = '0'). This means then when the circuit goes from capture to test mode, the flag becomes meaningful and its value designates whether the two bitstream are equal or not.

The *Output Enabler* permits the observation of the TestRes only after comparing the whole test vector. It is composed of a down counter with parallel load that loads the value #SFF whenever the scan operation is not enabled. Therefore, when the circuit goes to test mode, it start counting and after #SFF clock cycles its terminal count allows outputting the TestRes signal through the AND gate.

The *I/O Buffers* allow sharing the same pin for Sin and TestRes. The proposed solution requires, besides Scan-In and Scan-Enable signals (Sin and Sen), the Sexp signal, which replaces the standard Scan-Out signal, and the additional TestRes. However, Sin and TestRes are not used at the same time, therefore it is possible to use bi-directional buffers shared between them, as shown in Fig. 7.3. During the shift operation the pin can be set as input and used by the tester to feed the circuit with the input vectors, whereas during the capture operation the pin is activated as output to deliver the previous comparison result.

The secure comparator does not impact the fault coverage. In fact, each test response is compared to the expected one as in a classical ATE-based test scheme. Therefore, the achievable fault coverage is not altered. Test time is not increased either, since the expected responses are scanned-in at the same time as the next input vector is scanned-in.

Therefore, the proposed secure comparator allows similar diagnostic resolution as it can be obtained with the classical scan scheme. The only difference resides in the matching procedure between the obtained responses and those stored in the fault dictionary. In the classic scheme this is done off-line (i.e., after collecting all responses from the circuit), while in this case all potential faulty responses must uploaded on the DUT for comparison with the actual faulty response, thus requiring additional time. The diagnosis is however limited to only modeled faults [13].

7.4 Secure Test Access Mechanism

Solutions presented so far make the assumption that the classical scan chain approach is either not used (as in the functional test or BIST solutions) or partially used (as in the on-chip comparison approach). In this section, we analyze all solutions that consider the use of standard scan chains, with Scan-In, Scan-Out, and Scan-Enable signals.

The idea in all the following solutions is that the circuit can be either tested or used in normal mode (test mode vs. mission mode). In normal mode, the scan path is not supposed to be used for shifting out the device internal state, while in test mode the circuit can switch from scan mode (i.e., shift mode asserted by the Scan-Enable signal) to capture mode and vice-versa.

The issues that are solved in all proposed methods are the following:

- **How to start the test mode**. The test mode can be activated through an additional signal (besides Scan-Enable), or by using an authentication procedure. A password-based authentication method is proposed in [18]. In order to proceed with the standard test procedure, the user has to pass through an initial authentication phase with several steps. In each step, the user has to insert the test patterns, which contain a secret key that is compared to a golden one. Only if the user correctly guesses the different keys in all steps, the test session starts (and the scan output is observable). In [19], the scan chain contains k spare flip-flops. When the user enters a test pattern, the values in the k flip-flops are compared with a hardwired password. Only if all the bits match, then the response vector can be observed. Similar solutions (also applied to test standards) are described in [20, 21]. Moreover, in [22, 23], the authentication procedure can be restricted to some of the cores in the circuit.
  In [24] the authors use a finite state machine that observes the Scan-Enable signal. Whenever the Scan-Enable is asserted (i.e., a scan shift operation is started), the circuit goes automatically in test mode. The normal mode can be restored only by resetting the circuit.
- **What to do when switching from normal to test mode**. As shown in [25], by resetting all FFs when switching from normal mode to test mode, no secret information can leak. In [24] the authors propose to flush the content of the scan chain (by ANDing the Scan-Out with a control signal set to 0) for a number of clock cycles equal to the number of flip flops in the scan chain, during the first shift operation performed after switching from normal to test mode. The flushing

operation guarantees that any secret will be kept inside the circuit, and any further scan operation will not reveal then any confidential information.

- **What to do in normal mode**. When in normal mode, no shift operations on the scan chain have to be tolerated. Therefore, several solutions have been proposed to monitor the status of the scan chain in order to be sure that unauthorized shift operations are not performed, or to scramble the content and the order of the FFs so that, even in case of observation of the scan chain, no useful data can be obtained. The hypothesis in all these solutions is that the attacker manages to bypass the mechanism that allows switching from normal mode to test mode, thus controlling the shift operations.
  Scrambling countermeasures ensure the confusion of the stream shifted out from the scan outputs for unauthorized testers [26, 27]. In [28] the authors propose an obfuscation technique based on the implementation of nonlinear functions between two FFs in the scan chain, so that to alter the content of the scanned bitstream if a shift-out operation is performed.
  In order to obfuscate the content of FFs, the authors in [29] propose to add a latch for every FF in the circuit. During shift operations, the content of the FF is XORed with the content of the latch, which contains a past state of the FF. From the external, the scan structure seems to be changed time by time. In [30, 31], the authors propose to swap the position of the scan cells in a chain, by carefully selecting the proper FFs to be swapped.
  Concerning the intrusion detection (that would happen in normal mode), in [32] the authors propose to detect unauthorized scan enable settings. This technique consists of connecting some leafs and the root of the Scan-Enable tree to a comparator. When the authentication has been bypassed the Scan-Enable signals on every SFF is supposed to be disabled, therefore any illegal shift will raise an alarm by detecting that at least one observed Scan-Enable signal is active. The same authors propose to add "spy" FFs in the scan-chain to detect unauthorized shift at mission time. These spy FFs are inserted between actual SFFs and set to a constant value (for instance 0) in normal mode. Then, the outputs of these FFs are compared to check if they store the same value. Because illegal shifts will rapidly load these spy FFs with a different value, intrusion can be detected.
- **What to do in test mode**. When in test mode, no secrets should be delivered. Solutions have been proposed in order to mask the actual content of the secret key and to use a *shadow* key instead, used only for test purposes [24, 25].

7.5 Industrial Solutions

Semiconductor Industry develops secure Integrate-Circuits (IC) for many different application domains such as Banking, access enabler, e-government (ID, passport), medical, communication (mobile, wifi…), Internet of Things. For such products, security (i.e., Confidentiality, Integrity, Authenticity, Availability) has to be guaranteed during the whole lifetime of the product, from development to final application.

At design time, the security must thus be questioned when implementing Design-for-Testability features as well.

As detailed in the previous chapter, classical DfT approaches such as scan design jeopardize data confidentiality, and thus, dedicated DfT solutions and structures have to be used to ensure compatibility between security and testability. Standard-DfT weaknesses are discussed from an industrial point of view in Sect. 7.5.1. Section 7.5.2 focuses on industrial constraints when designing a Secure DfT solution. Section 7.5.3 aims at describing generic secure DfT solutions able to be used in a standard design flow. Section 7.5.4 presents DfT solutions for the particular case of memories (both RAM and ROM).

7.5.1 Standard DfT Weaknesses

In order to propose efficient secure DfT solutions, "standard" DfT practices have to be analyzed to better understand potential weaknesses against attacks. In previous chapters, several attacks have been described showing how DfT structures could be reused during attacks. As for any real industrial threat, we also need to identify attackers to better understand their own objectives and capabilities in order to build adapted protections.

Attackers are classified in the following way (as shown in Fig. 7.4):

- *Competitors*, whose main objectives are reverse engineering and IPs cloning. We assume they have high-level competencies in DfT techniques and a utilization of up-to-date failure analysis tools. We can also assume knowledge from former employees;

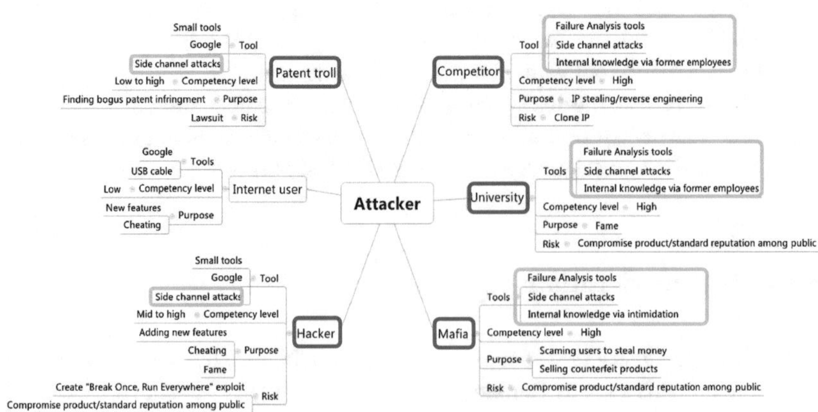

Fig. 7.4 Taxonomy of DfT attackers

- *Patent troll*, whose main objective is to demonstrate patent infringement. Competencies in DfT are considered medium and they generally do not have access to advance design analysis tools;
- *Academics*, whose main objectives are technical challenges resolution and fame. Academic competencies in DfT are generally high and it is reasonable to consider a full access to up-to-date failure analysis tools. We can also assume knowledge from former employees;
- *Hackers*, whose main objectives are either cheating end user applications or adding new features/backdoors. Their competencies in DfT are low to medium and they generally do not have access to advanced IC analysis tools;
- *Mafia*, which have as main objective to earn money by stealing end user private information or by selling counterfeit products. Mafia could have a high level of competency and access to failure analysis tools. We have to assume that mafia has also access to IC internal knowledge via employees' intimidation.
- *Internet users*, who are able to perform only software attacks, do not target DfT logic as line of attack. Indeed they generally do not have DfT techniques background and fewer more DfT interfaces cannot be directly controlled by an internet-controlled functional interface.

Based on the above-described characteristics of possible attackers, we can derive general recommendations for a Secure DfT:

- Former employees: Secure DfT should respect Kerckhoff's principle, i.e., full knowledge of the protection implementation is not enough to "open" the system, data integrity, and confidentiality must rely on a cryptographic key, not on the knowledge of the cryptographic algorithm or its hardware implementation;
- Fault Analysis tools: Secure DfT should implement countermeasures to test-based attacks in such a way that defense might not be disabled by tools, such as microprobing, focused ion beam (FIB) or laser beams;
- Side-Channels Attacks: Secure DfT implementation should take care of information leakage.

General recommendations, from which we could extrapolate few specifications for a Secure DfT implementation strategy, are listed in the following:

- Secure DfT should embed a lock mechanism to avoid a direct access to test infrastructures in any operational mode (test and mission modes);
- Secure DfT should avoid facilities for free switching between mission and test modes;
- Secure DfT should protect memories contents by avoiding direct read and write accesses;
- Secure DfT should protect critical sequential elements (flip-flops, latches) contents during test mode to avoid reverse engineering and application data extraction.

Secure DfT Techniques able to fulfill industrial constraints will be described in following chapters.

7.5.2 Secure DfT and Industrial constraints

Several secure DfT solutions have been already published, but almost all proposed academic solutions target a single attack mechanism while real products have to be protected against several types of attacks. Moreover, proposed secure DfT solutions may not be supported by standard design tools used in IC industry and implementation issues may lead swiftly a secure solution that looks good on paper to turn to a nightmare on real implementations. From an industrial point of view, a "good" solution must thus be easily implemented thanks to computer-aided-design (CAD) tools. In addition to the impact on the development time, the silicon overhead is another very important criteria that also has to be considered.

To summarize, an effective industrial secure DfT solution should guarantee a high level of security in parallel to fulfill several industrial constraints:

- *Quality*: Secure DfT must allow to achieve a high test coverage for targeted fault models (Stuck-at, Transition, Bridge…) by reusing as much as possible existing tools capabilities, such as scan insertion and automatic test pattern generator (ATPG);
- *Low Cost*: Secure DfT silicon overhead as well as manpower needed to implement the solution, must be part of the criteria to select an industrial solution;
- *Time-To-Market*: The secure DfT flow must be inserted in an automated design flow and its capability to be developed as a generic solution are important criterion for an industrial Secure DfT solution.

Moreover, it is important to also consider the complete IC manufacturing process during the DfT secure solution evaluation. Indeed, only a few Semiconductors Companies are able to build an IC from design to final customer deliveries. Most of the time, third part companies are used to execute parts of the manufacturing flow, such as mask manufacturing, IC manufacturing, wafer test, wafer grinding and sawing, or final manufacturing test. In these cases, it is important to review the complete security concept, including DfT, to ensure that using a third company to execute one or several manufacturing steps does not break the security concept, by creating a weakness that could be used by one of the attacker defined previously. Generally, confidentiality contract and details processes are defined between IC Company owner and third part companies to reduce information leakage. Such processes are also internally used to avoid leakage by employees.

7.5.3 Industrial-Constraint-Aware Secure DfT

Manpower resources, time-to-market constraints and incompatibility with industrial DfT tools often avoid a direct implementation of state-of-the-art and academic secure DfT solutions. The following subchapters aim at describing generic secure DfT solutions able to be used in a standard design flow.

Before going into details of secure DfT solution, it is important to note that for a real product, security is never related to a single secure structure. Several secure solutions are put all together, as several layers, in order to build at the end a robust secure solution able to protect IC against several attack types. Therefore, while each individual solution could be assessed as being not as secure against dedicated attack as a single solution, it must be kept in mind that high security level is reached when all security layers are implemented.

7.5.3.1 Test interface

In a bank office, security relies on the safe more than into the front door! This example also applies to industrial test interfaces. The test interface is the main door for all standard usages of the test structures, and it must be controlled at the different stages of the product life:

- At die level, for wafer test;
- At package level, for final test;
- During product life, to analyze customer returns.

Except during these steps, the test interface must be locked, especially at end user site, to avoid any unauthorized usage. Moreover, several solutions can be used at the same time to protect the test interface. In industrial solutions the following recommendation are often used:

- Test interface must not use dedicated pad, but must share functional pads to avoid simultaneous usage between functional and test mode;
- Test interface must not use a standard communication protocol to avoid usage of commercial tools (probes) to perform complex automatics attacks;
- Test interface must not be easily enabled, to avoid fast switch between test and functional mode.

From a simple magic sequence detection, used to unlock the test interface, up to a complex challenge-response authentication mechanism, those three points above are referring design solutions for which each company should develop their own solutions. It is important to notice that many test interface lock structures, are relied on embedded non-volatile contents, which are by construction not known before the first access. So we can assume that product security will start after the first test, in which non-volatile elements are initialized.

Several structures can be used to detect wafer test phase, before to be permanently modified for the next production flow step. The most common structure are:

- OTP (One Time Programmable) memory, Poly Fuse and Laser Fuse. Their implementation require an important area overhead, and they are easy to localized and repaired;
- MTP (Multiple Time Programmable) memory, i.e., a non-volatile memory which needs to be protected to avoid abnormal write access;

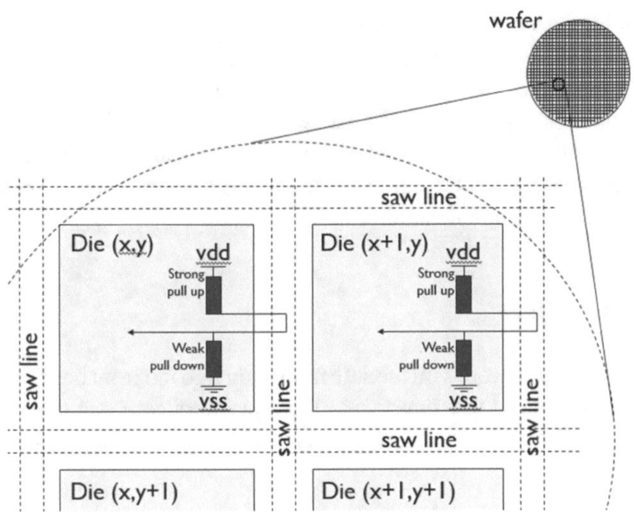

Fig. 7.5 Saw Bow used to detect wafer step during test

- Saw bow (see Fig. 7.5), i.e., an electrical connection based on strong pull-up and weak pull-down, which are physical interconnected by a metal line across the saw line of the wafer. Strong pull-up resistance will set the value before sawing the wafer, while the weak pull-down resistance will set the opposite value after the saw. Such a technique must be compatible with the production flow and technology.

Those wafer-test-step detection techniques can be associated to additional structures to build a secure test interface using different access mechanism during wafer test or final test. In the example of Fig. 7.6, a password (1) has to be sent during the first clock cycles after the reset (2) to enable the test interface, if wafer-test step is detected (3). During wafer test or during the sawing, the structure used to detect wafer level test is permanently disabled, then for all next test steps, a functional authentication, which could use an embedded encryption module, is required to enable the test interface.

In order to avoid the usage of wafer test detection, a possible solution is based on the challenge–response technique. In this case, the same test entry sequence can be used for all test steps. The basic principle consists in tester authentication. For example, the DUT can generate a random number and it encrypts this number with a private key (asymmetric encryption). The encrypted random number is sent to the tester, decrypted using the public key, then sent back to the DUT to be compared with the original one. The result of the comparison will be used to enable or disable the test interface. The main drawback for such implementation is the amount of functional logic needed to perform the complete test entry sequence. Moreover, if a fault affects the authentication module, it will be impossible to perform diagnosis (i.e., fault localization).

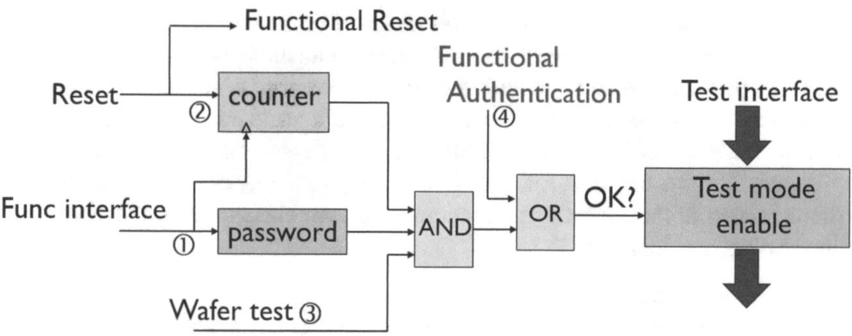

Fig. 7.6 Secure test interface

7.5.3.2 Test Control Register

Scan based design generally requires the implementation of extra test registers to control clocks, resets, clock-gating structures, and analog interfaces. A test register is considered as a standard design implementation without CAD tools limitation. Because these test registers are good candidates for attack based on fault injections, or to takeover scan chains control using micro-probing, some secure structures have to be implemented for these registers. Basic security recommendation are based on redundancy and control:

- Test control registers must be forced in reset state as long as test interface is locked
- Test control registers must be "write-only" registers
- Test control registers must be protected against fault injections by using Error-Correcting Code (redundancy bits) to detect invalid register values for instance.

A multitude of design solutions exists to secure usage of a test control registers using above rules; each company could apply its own techniques.

7.5.3.3 Secure Scan Test Structure

Scan test use a structured methodology supported by commercial tools:

- Scan Design: Scan chain creation (scan insertion) is often embedded into the synthesis tools (e.g., Cadence RTL Compiler [33] or Synopsys Design Compiler [34]) or sometime delivered as a standalone tool (e.g., Mentor Graphics Tessent Scan [35]);
- Test Pattern Generation: test stimuli applied through direct scan chains access or dedicated test decompression structures are automatically generated thanks to Automatic Test Pattern Generator (ATPG) tools (e.g., Synopsys TetraMAX [36], Mentor-Graphics TestKompress [37], Cadence Encounter [38]).

Complexity of nowadays ICs requires usage of those automatic tools to rapidly reach an acceptable test efficiency level. A default scan implementation will however result in a poor implementation with regard to the security of the data processed by the circuit. On the other side, any deviation from a standard scan structure could result to a CAD tool blockage. In following paragraphs, we will describe several tool-compatible secure scan test solutions. These solutions are classified into four groups related to the scan chains structure, the scan chains usage, the scan chain compression, and logic BIST.

Scan chains structures
Utilization of the scan chains to identify registers content or to perform reverse engineering operation is simplified when the scan chain structure is known, i.e. when the hacker knows the position of each scan flip-flop in the chain. This scan chain structure can be easily identified when using a standard scan insertion flow. Scan insertion tools are indeed using by default the flip-flop alphabetic order (including design hierarchy) to interconnect registers within scan chains. For example the 128 bit AES key flip-flops are most probably all connected together from bit 0–128 within the same scan chain (in reality 0, 1, 10, 100, 101, 102, …, 11, 110, 111, …, 12, 120, 121, …, 13, 14, …, 2, 20, 21, 22, …, 30, …, 99).

In order to improve the scan chains security, a proposal consists in scrambling the scan path by using a standard scan reordering flow for which only the functional paths constraints are used for place and route, as shown in Fig. 7.7. Scan interconnection will be then based on flip-flop physical location (interconnected flip-flops which are close to each other). The first drawback of this approach is that this technique could generate a lot of hold timing violation that will have to be corrected. Anyhow, this flow is supported by all scan insertion tools, and will perform a kind a random local scan chains scrambling. In addition as demonstrated in paper [39], insertion of several inverters on scan shift path will improve robustness against reverse engineering

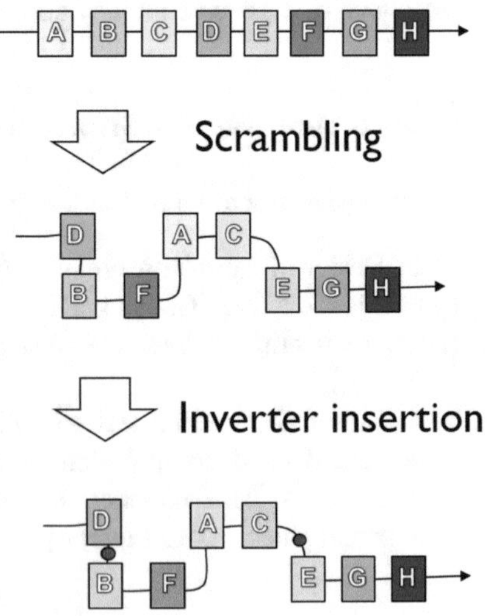

Fig. 7.7 Scrambled scan registers

attacks. After the scan insertion step, you could randomly insert inverter on scan shift paths (between scan flip flops), using netlist manipulation commands available in synthesis or scan insertion tools. Those static inverters will be fully transparent for the ATPG flow, which is already able to deal with scan chain inversion. Please note that dynamic inverter as proposed in papers [40, 41] are not supported automatically within an ATPG flow. The second drawback of this approach is that differential scan chain analysis [41] allows to deal with unknown scan chain implementations and to perform scan chain attacks anyway at the cost of longer computation time.

Scan chains usage
One of the most common scan chain attack is based on the utilization of the scan chain to dump flip-flops values through the scan output thanks to the scan shift mode. Attacks happen during a functional execution, which is stopped at a critical point to dump, for example, data related to the secret key in a crypto-coprocessor. Micro-probing techniques could be used to locally force the scan shift mode or to hack the main test controller to control the whole scan chains. In order to improve the scan test security, an efficient proposal is to insert sensors to detect abnormal scan shifts, i.e. scan shift operations during mission mode. Several sensors are proposed in the literature, but real implementation is not always easily feasible.

First proposal is to monitor the Scan-Enable signal at different locations in the design, in order to detect a local shift mode by observing a value change on one of the observed node [42], as shown in Fig. 7.8. Unfortunately, this proposal is not so easy to implement due to the fact that Scan-Enable signal is automatically connected to all flip-flops during the scan insertion phase and buffer-tree append only during Back-End phase which is normally too late to perform netlist manipulation.

A second technique based on the same basic principal is using spy flip-flops [42] instead of direct observation of the Scan-Enable signal, as shown in Fig. 7.9. The spy flip-flops' D inputs are tied to a fixed value, while there are inserted in the scan chains at random position. All spy flip-flop's Q outputs are observed and compared to a central Scan-Enable signal. All those spy signals must be fixed during mission mode, thanks to the tied flip-flop' D (functional) inputs, while they will toggle during scan shift mode allowing to detect an abnormal (local) scan shifts. It is possible to implement this solution in a generic way, by creating a single "spy flip-flop" as a sub module (hard macro) instantiated several times in the design to protect. Spy flip-flops can then be spread over the scan chains using dedicated scan insertion constraints command.

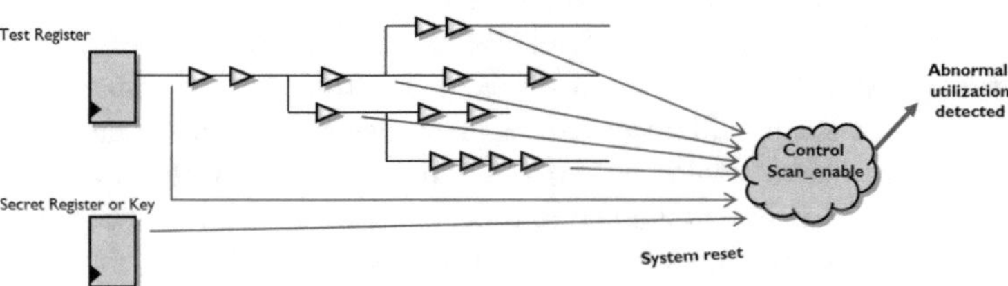

Fig. 7.8 Scan-enable control

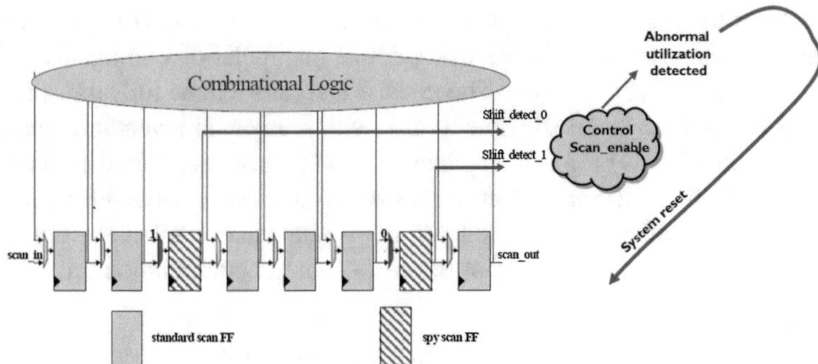

Fig. 7.9 Spy scan-registers

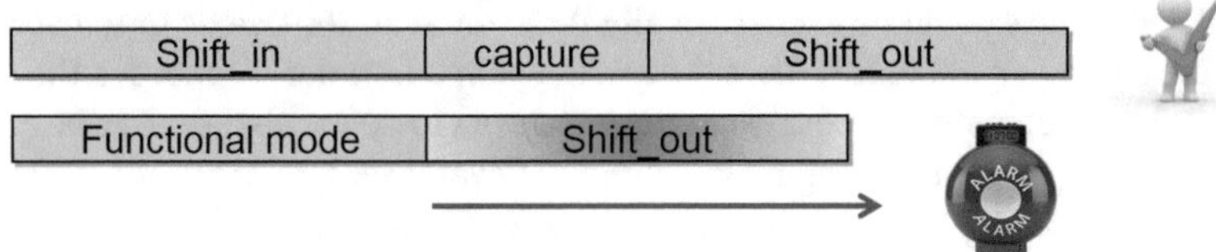

Fig. 7.10 Scan shift input detection

A third solution to avoid observation of FFs' functional values is to erase all functional values before allowing any scan shift operation. As described in the previous section, a generic solution is based on the utilization of a sensor able to detect the scan mode (scan shift or scan capture) [24]. Scan mode sensor is a simple counter (not "scanned") reset thanks to the scan shift enable signal and clocked thanks to the scan clock signal, as shown in Fig. 7.10. This counter counts scan shifts until a predefined value, i.e. the number of flip flops in the longest scan chain, before enabling the scan shift out. So, scan chain outputs are blocked until scan inputs values are replaced all functional internal values. Meaning that for a standard scan usage, starting by a complete scan shift input before a first scan shift output this implementation is fully transparent, while switching from functional mode to scan shift mode is blocked. This sensor implementation is fully transparent for ATPG or during standard scan test execution.

A fourth solution is based on a kind of scan pattern watermarking. The solution described in [43] uses a shift register (few bits) inserted at the end of one scan chain which have to receive a predefined watermark (hardcoded value) to enable the scan shift output paths, as shown in Fig. 7.11. Inserting the watermark register at the end of a scan chain, will ensure a complete scan shift input sequence has been executed, before allowing any scan shift out. It has a similar objective as the previous solution; functional values contained in flip-flops are erased by the scan shift input needed to load the watermark code. The watermark value is validated at the end of each scan shift input by the scan-enable signal transition (from scan shift to scan

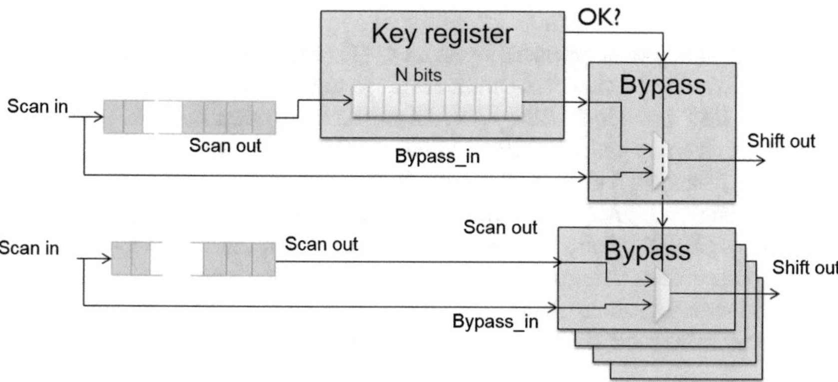

Fig. 7.11 Scan watermark detection

capture), in order to validate the following scan shift out. Regarding implementation, the watermark shift register, has to be scan inserted at the end of the longest scan chain using scan insertion constraints. Then during ATPG, the watermark code can be easily forced using ATPG constraints, which allow to predefine scan capture flip-flop values, even through a scan compression structure.

Scan chains compression
Scan compression structures proposed by standard ATPG tools, provide security, if well implemented, by applying a one-way function on information. Compression structure is basically reducing direct scan chain inputs controllability and scan chain outputs observability, with a minimum coverage impact. However, scan compression structure used alone is not sufficient to guaranty efficient security as demonstrated in [41], but associated with others secure scan solution as proposed above, it could bring security benefit. Anyhow, in order to be useful for security, test compression structures have to follow some rules regarding their implementation. First it is important to notice that the 3 main ATPG providers, Cadence, Mentor-Graphics and Synopsys, are using XOR tree as scan output compactor associated to X making structure. XOR tree structure is linked to the selected compression ratio. You have to select a big enough compression ratio, in order to always have several scan chains combined together to generate one XOR tree outputs. Author experiments show that output compression ratio of 20, will guarantee a minimum combination of 3 or 4 scan outputs to generate each XOR tree outputs. Moreover, special care about X masking logic functionality should be paid, in order to avoid any setting able to reduce the number of scan chain combination through the XOR compactor. Indeed, some ATPG provide dedicated X masking mode, in which a single scan is observed at the time, to improve observability (test coverage). Those modes must be carefully removed from the design implementation in order to use scan compression structure as security element. After such modifications, you may have to use dedicated ATPG commands to avoid usage of those special X Making modes by ATPG during test stimuli generation.

Logic BIST
Logic BIST solution as proposed by main CAD tool providers is really attractive for a security point of view. Few commands are send from outside and as response the Logic BIST provide PASS / FAIL information. No data leakage, scan chains are loaded and compared internally. Moreover, BIST can be executed in application to ensure IC integrity before each usage, feature that could be part of the security strategy. Main problems for an industrial point of view are the silicon overhead required by LBIST solutions, especially to test a complete IC, and the fact that LBIST solution does not provide diagnosis capability. So recommendation is not to use LBIST as a solution to protection internal data during production test, but as a test solution to be used in application for an integrity test.

7.5.4 RAM/ROM Test

In secure ICs RAMs and ROMs, contents must be protected against unauthorized read-out and no functional data must be directly writable. For all RAMs embedded in a secure IC few DfT rules must be guaranteed:

- Test sequence must start by a complete erase of the functional value;
- RAMs access must be disabled during scan test mode, to avoid data read-out through scan chains;
- Direct access to RAMs contents from pad level must be avoided.

In order to fulfill those secure requirements, a classical solution is based on the use of standard March-Test algorithms. March-Test is actually the most common test method used within semiconductors industry to test RAM. In order to secure the memory, March-Tests must be associated to isolation structures able to avoid RAM access during scan test mode.

On the other side, ROMs embedded in a secure IC must also follow few DfT rules in order to protect their contents against abnormal read-out operations:

- Direct access to ROMs contents from pad level must be avoided;
- ROMs access must be disabled during scan test mode, to avoid data read-out through scan chains.

In order to fulfill those secure requirements, standard ROM BIST associated to a MISR (Multiple-Input Signature Register) can be used, to create a signature based on ROM contents. Such a structure is a standard ROM test solution within semiconductor industry.

7.6 Conclusions

DfT and security are often presented as antagonist, indeed the DfT has as objective to maximize internal nodes observability and controllability, while security constraints do not allow any internal information to be controlled or extracted. However, we showed that DfT can handle security constraints while being supported by commercial tools. We described secure scan test solution and secure RAM/ROM test techniques. Pro and cons of the insertion of such techniques in a design flow are also being discussed.

To conclude, DfT and Security could be compatible if carefully implemented.

References

1. Abramovici M, Breuer MA, Firedman AD. Digital system testing and testable design, Revised Printing, IEEE Press; 1990. ISBN 0-7803-1062-4.
2. Richard A. Wheelus TD, Haverkos KWJ, Integrated circuit memory using fusible links in a scan chain. U.S. Patent US5677917, issued April; 1996.
3. Bardell PH, McAnney WH, Self-testing of multichip logic modules. In: Proceedings of international test conference; Nov. 1982. p. 200–04.
4. Schubert A, Anheier W. On random pattern testability of cryptographic VLSI cores. J Elect Test Theory Appl. 2000;16(3):185–92.
5. Shannon C. A mathematical theory of communication. Bell Syst Tech J. 1948;27(4):379–423.
6. Shannon C. Communication theory of secrecy systems. Bell Syst Tech J. 1949;28(4):656–715.
7. Feistel H. Cryptography and computer privacy. Sci Amer Mag. 1973;228:15–23.
8. Di Natale G, Doulcier M, Flottes ML, Rouzeyre B. Self-test techniques for crypto-devices. In: IEEE transaction on VLSI systems, vol. 18, Issue 2. p. 1–5, Feb 2010. DOI:10.1109/TVLSI. 2008.2010045.
9. Doulcier M, Flottes ML, Rouzeyre B. AES-based BIST: self- test, test pattern generation and signature analysis. In: Proceedins of 4th IEEE international symposium electron design, test applications (DELTA), 2008. p. 314–21.
10. Joan D, Vincent R. The design of rinjael, AES—the advanced encryption standard. 2nd ed. New York: Springer.
11. Recommendation for the Triple Data Encryption Algorithm (TDEA). Block Cipher, Special Publication 800–67, Gaithersburg, MD: National Institude Standards Technology (NIST); 2008.
12. Karaklajic D, Kneževic M, Verbauwhede I. Low cost built in self test for public key crypto cores. In: Workshop on fault diagnosis and tolerance in cryptography (FDTC). Santa Barbara, CA. 2010. p. 97–103. doi:10.1109/FDTC.2010.12.
13. da Rolt J, Di Natale G, Flottes ML, Rouzeyre B. Thwarting scan-based attacks on secure-ICs with on-chip comparison. IEEE Trans Very Large Scale Int Syst. 2014;22(4):947–51. doi:10. 1109/TVLSI.2013.2257903.
14. Sudeendra Kumar K, Lodha K, Sahoo SR, Mahapatra KK. On-chip comparison based secure output response compactor for scan-based attack resistance. In: 2015 international conference on VLSI systems, architecture, technology and applications (VLSI-SATA). Bangalore; 2015. p. 1–6. DOI:10.1109/VLSI-SATA.2015.7050467.
15. Talatule SD, Zode P, Zode P. A secure architecture for the design for testability structures. In: 19th international symposium on VLSI design and test (VDAT). Ahmedabad. 2015:1–6. doi:10.1109/ISVDAT.2015.7208090.
16. Wu Y, MacDonald P. Testing ASICs with multiple identical cores. IEEE Trans Comput Aided Des Int Circ Syst. 2003;22(3):327–36.
17. Poehl F, Beck M, Arnold R, Rzeha J, Rabenalt T, Goessel M. On-chip evaluation, compensation and storage of scan diagnosis data. IET Comput Dig Tech. 2007;1(3):207–12.
18. Paul S, Chakraborty R, Bhunia S. VIm-scan: a low overhead scan design approach for protection of secret key in scan-based secure chips. In: Proceedings of 25th IEEE VLSI test symposium, May 2007. p. 455–60.

19. Lee J, Tebranipoor M, Plusquellic J. A low-cost solution for protecting IPs against scan-based side-channel attacks. In: Proceedings of 24th IEEE VLSI test symposium, May 2006, p. 1–6.
20. Novak F, Biasizzo A. Security extension for IEEE Std 1149.1. J Elect Test. 2006;22(3):301–3.
21. Chiu G-M, Li JC-M. A secure test wrapper design against internal and boundary scan attacks for embedded cores. IEEE Trans Very Large Scale Integr Syst. 2012;20(1):126–34.
22. Wang X, Zheng Y, Basak A, Bhunia S. IIPS: infrastructure IP for secure SoC design. IEEE Trans on Comput. 2015;64(8):2226–38. doi:10.1109/TC.2014.2360535.
23. Dworak J, Conroy Z, Crouch A, Potter J. Board security enhancement using new locking SIB-based architectures. In: IEEE international test conference (ITC), WA: Seattle; 2014. p. 1–10. doi:10.1109/TEST.2014.7035355.
24. Da Rolt J, Di Natale G, Flottes ML, Rouzeyre B. A smart test controller for scan chains in secure circuits. In: Proceedinigs IEEE 19th IOLTS, July 2013. p. 228–9.
25. Yang B, Wu K, Karri R, Secure scan: a design-for-test architecture for crypto chips. IEEE Trans Comput Aided Des Integr Circ Syst. 2006;25(10):2287–93.
26. Hely D, Flottes ML, Bancel F, Rouzeyre B, Berard N, Renovell M. Scan design and secure chip [secure IC testing]. In: Proceedings of 10th IEEE IOLTS, July 2004. p. 219–24.
27. Lee J, Tehranipoor M, Patel C, Plusquellic J. Securing scan design using lock and key technique. In: Proceedings of 20th IEEE international symposium DFT VLSI system, Oct. 2005. p. 51–62.
28. Fujiwara H, Fujiwara K. Strongly secure scan design using generalized feed forward shift registers. IEICE Trans Inf Syst. 2015;E98-D(10):1852–55.
29. Atobe Y, Shi Y, Yanagisawa M, Togawa N. Dynamically changeable secure scan architecture against scan-based side channel attack. In: International SoC design conference (ISOCC). Jeju Island; 2012. p. 155–8. doi:10.1109/ISOCC.2012.6407063.
30. Ali SS, Saeed SM, Sinanoglu O, Karri R. Novel test-mode-only scan attack and countermeasure for compression-based scan architectures. In: IEEE transactions on computer-aided design of integrated circuits and systems. 2015;34(5):808–21. doi:10.1109/TCAD.2015.2398423.
31. Saeed SM, Ali SS, Sinanoglu O, Karri R. Test-mode-only scan attack and countermeasure for contemporary scan architectures. In: IEEE international test conference (ITC), Seattle, WA; 2014. p. 1–8. doi:10.1109/TEST.2014.7035357.
32. Hely D, Bancel F, Flottes ML, Rouzeyre B: Secure scan techniques: a comparison. In: Proceedings 12th IEEE ISOLT, Jan. 2006. p. 119–24.
33. http://www.cadence.com/products/ld/rtl_compiler/pages/default.aspx
34. http://www.synopsys.com/Tools/Implementation/RTLSynthesis/DesignCompiler/
35. https://www.mentor.com/products/silicon-yield/products/scan
36. http://www.synopsys.com/Tools/Implementation/RTLSynthesis/Test/Pages/TetraMAXATPG.aspx
37. https://www.mentor.com/products/silicon-yield/products/testkompress/
38. http://www.cadence.com/products/di/edi_system/pages/default.aspx
39. Yang B, Wu K, Karri R. Scan based side channel attack on dedicated hardware implementations of Data Encryption Standard. In: International test conference, 2004. p. 339–44.
40. Nara R et al. RScan-based attack against elliptic curve cryptosystems. In: ASP-DAC, 2010. p. 407–12.
41. Darolt J, Di Natale G, Flottes ML, Rouzeyre B. Are advanced DfT structures sufficient for preventing scan-attacks?. In: VLSI test symposium, 2012. p. 246–51
42. Hely D, Bancel F, Flottes M-L, Rouzeyre B. Securing scan control in crypto chips. JETTA. 2007;23(5):457–64.
43. Pugliesi-Conti PH. Circuit for securing scan chain data, patent filed, March 25, 2011, Publication number: 20120246528.
44. van de Goor AJ. Testing semiconductor memories: theory and practice. John Wiley and Sons, 1991.
45. Zarrineh K, Upadhyaya SJ, Chakravarty S. A new framework for generating optimal march tests for memory arrays. In: IEEE international test conference, 1998. p. 73–82.

Chapter 8
Malware Threats and Solutions for Trustworthy Mobile Systems Design

Jelena Milosevic, Francesco Regazzoni and Miroslaw Malek

8.1 Introduction

Rapid adoption of mobile devices and their increased usage to perform financial transactions and to send or store sensitive information, attracted the attention of criminals and all sorts of trouble makers and increased their interest in tampering with these devices to gain profit, to collect private and sensitive data, or simply to cause malfunctioning. To guarantee the security of a mobile device, it is necessary to provide it with robust and trusted hardware. Trusted hardware means that the used components should not contain hardware Trojans, which are malicious modifications of the underlining hardware in order to access maliciously the target device. Robust hardware means being resistant against physical attacks.

Being mobile and widely present, mobile devices can get into physical possession of the attacker, which makes them prone, as the large majority of other embedded systems, to threats caused by physical attacks. Physical attacks are attacks which aim at gaining access to sensitive information by exploiting the physical leakage of the implementation of security primitives. The most notable example of these attacks is by using power analysis [22], where the secret key is extracted by analysing the dependency of the power consumed by the device and the secret data being processed. However, using power analysis is not the only physical attack which exists. There are also other methods that may exploit timing and that were successfully used in the past: timing difference [21], electromagnetic emissions [29], and deliberate fault injection [19].

However, state-of-the-art mobile devices are not composed of solely hardware (including several cores and dedicated accelerators like GPUs) but also a plethora of software. For this reason, to guarantee the overall robustness of the device, it is not sufficient to protect only the hardware but also software routines have to be trusted,

J. Milosevic (✉) · F. Regazzoni · M. Malek
Advanced Learning and Research Institute, Università Della Svizzera Italiana, Lugano, Switzerland
e-mail: jelena.milosevic@usi.ch

N. Sklavos et al. (eds.), *Hardware Security and Trust*,
DOI 10.1007/978-3-319-44318-8_8

since the number of security breach caused by software is significantly growing. This is indeed visible in the reported number of malicious software or shortly malware, which is increasing very fast. According to [6], the total number of mobile malware samples grew 17 % in the second quarter of 2015. A different source [3], states that currently about three over four applications in China are malware.

Malware is software deliberately created to harm the device where it will be executed. Some of the effects which malware can have are stealing of sensitive information, the possibility of taking control of the overall operation of the system, and the damaging till the complete disruption of the device.

The number of encountered attacks on mobile devices is growing, so as the number of malware samples and malware families. With increased number of mobile families, also the behaviour of malware is changing, progressing and becoming more difficult to detect. Under these increasingly difficult circumstances, the detection algorithms have to cope with the variety of malicious behaviour, and be able to provide an effective detection, without generating an amount of false positives that would disturb users. The way to cope with it, and provide an effective solution, is mostly, by increasing effectiveness of algorithms that in turn may require higher complexity and taking into account more parameters about the system. However, mobile malware detection systems have to be run in resource-constrained and battery-operated environments that neither have the computational power to run extremely complex algorithms nor can support algorithms that drain the battery too quickly. Finding an effective detection algorithm, that is at the same time suitable for battery-operated mobile devices, is a challenging task.

In view of the increasing relevance that this problem has in mobile devices, and considering the effect of malware in the whole trustworthiness of a system, this chapter surveys existing mobile malware detection threats and proposed solutions and sketches main research trends. The main goal of the work is to evaluate current approaches with respect to the effectiveness of the solution, and its consumption of resources.

8.2 Threats in Mobile Devices

For the second consecutive year, mobile devices are perceived as IT security's weakest link [1]. The threats, that were previously mostly concern of governments, financial institutions, and security vendors, are becoming more relevant in small enterprises and in personal lives [6]. The focus of Internet security is shifting from the desktop and the data centre to the home and Internet of Things, the pocket, the purse, and, ultimately, devices and infrastructure of the Internet itself [4].

The most severe threat that can affect mobile devices is malware. It is being able to completely damage a device or enable further attacks on the device that can perform unwanted actions. Most present malware types in mobile devices are rootkits, ransomware, bots, financial malware, logic bombs, viruses, worms and Trojan horses.

- **Rootkits** are a type of malware that is able to access parts of software for which regularly it does not have privileges. The access to privileged area is usually enabled by performing an attack on the system, either by exploiting systems vulnerabilities or guessing user's passwords. Once the attacker has the access to the root privileges of the system, the system is practically under full control of him or her and is prone to further manipulation. Due to this, rootkit detection is a challenging task, and sometimes the only way to cope with it is the replacement of the operating system.
- **Ransomware** is malware that locks the content of the user's device, and then asks the user to pay money, ransom, in order to enable normal usage. There are different ways to perform such attack on the system, starting from locking the screen of a device, or by using fake anti-virus software that, once installed on user's device, would prompt the message that the device is under attack and ask for money in order to remove the discovered infection. More advanced ransomware encrypts the data stored on the device and asks for money in order to provide the decryption key. In the last few years an increased number of ransomware attacks was recorded. More in detail, in the second quarter of 2015 their number increased by 58 % comparing to the first quarter of the same year [6].
- **Bots** are self-propagating malware with the goal to infect host machine and later connect to a server, bot master, and follow the obtained orders from it. Botnet is a network consisting of many host devices infected with bots, being available to perform Denial of Service attacks, send spam messages or simply enable further infections on host devices. Additionally, bots collect information from host devices and send it to the bot master. The collected information can be related to private user's data, financial transactions, user passwords, etc. Botnets, that until recently were mostly related to personal computers, since 2010 also attack mobile devices. One example of mobile bots with a goal to propagate malware is Plankton that appeared in 2011 and currently has more than 2000 different variants. More information about Plankton can be found in [37].
- **Financial Malware** has a goal of collecting accounts credentials and sending them to the attackers. Current Android malware can intercept text messages with authentication codes from customer's bank and forward them to attackers. Also, fake versions of legitimate banks mobile applications exist, hoping to trick users into giving up account details. Number of encountered attacks related to financial malware is increasing. This can be especially seen in the increase of banking malware, which attacks online banking customers. According to [5], number of encountered banking attacks increased from 71 to 83 % from first to second quarter of 2015.
- **Logic Bombs** are pieces of code intentionally inserted into a software system that set off a malicious function only when specified conditions are met. When activated, a logic bomb can perform different actions: display spam messages, delete or corrupt data, execute pieces of malicious code or have other undesirable effects.
- **Viruses** are type of malware that propagates by inserting themselves into another program and spreading together with it. The level of severity of viruses can vary from low, for example corrupting some files on the system, to very severe that

can disable and completely damage the operating system. Viruses are spreading together with the program they are attached to. It can happen by using Wi-Fi network, Bluetooth, message or email attachments.
- **Worms**, as opposed to viruses that depend on a host program to spread itself, operate more independently of other files. Still, same as viruses they are able to self-replicate and spread. In mobile devices, worms spread without user's knowledge, by using existing communication channels: SMS, MMS, and Bluetooth. First mobile malware, Cabir, that appeared in 2004, was a worm developed for Symbian operating system and ARM architecture that was able to spread itself via Bluetooth. Since then different variants of worms exist in mobile devices, causing users information leakage, disruption of services or sending premium rate messages.
- **Trojan Horses** (**Trojans**) are type of malware that appears as a legitimate software, but actually has malicious intents. Also, they are able to open a backdoor in a system, thus enabling further attacks. Due to their similarity with legitimate applications, detection of Trojans is a challenging task. At the same time, they are one of the most present malware types in mobile devices, especially devices running on Android operating system. One of the most famous is Spitmo, a Trojan which steals information from the infected smartphone, monitors and intercepts SMS messages from banks and uploads them to a remote server [37].

Apart from malware, threats that can also appear in mobile devices are classified as grayware or madware. According to [4], out of the 6.3 million apps analysed in 2014, one million were classified as malware, while 2.3 million were classified as grayware. A further 1.3 million apps within the grayware category were classified as madware.

- **Grayware** are all the programs that do not contain viruses and are not obviously malicious but that can be annoying to the user, like for example adware (advertising-supported software), that automatically delivers advertisements.
- **Madware** consists of different aggressive techniques developed in order to place advertisement in mobile devices, for example photo albums and calendar entries and to push messages to notification bar.

Apart from the listed threats, there are various other forms of malware, grayware, and madware that have different names and different forms. Some examples are *freeloading* that uses other people's phone by "freeloader" without permission of the user, *phishing* is looking for someone to get "hooked" and load malware/grayware or madware, and *spoofing* is pretending to be someone else (e.g. user's bank), win trust and exploit the credentials. Although the number of all possible threats that can happen in mobile devices is much higher, in this chapter we focus on and discuss in more detail the ones related to mobile malware, since it is currently the threat that can cause the most severe damage to devices, and particularly the ones that currently exist and are widespread in devices running Android Operating System, since the Android OS is currently the most used OS for mobile devices. With growing number of devices the complexity of systems is increasing, causing even more security threats to appear. It is estimated that the number of connected devices will continue to grow both in volume and variety, and that by 2020 it may reach 200 billion [2]. Although we have already seen attacks in ATMs, home routers, cars and medical equipment, these are just beginnings of attacks on IoT [6]. Most of these devices connect via Bluetooth that is known to suffer from many security flows, as stated in [2]. Apart from being able to collect data stored on these devices, attackers can also abuse their connections to smartphones. Symantec in [4] discovered that 20 % of applications related to health sent personal information, logins, and passwords over the wire in clear text.

The expansion of existing attacks is expected in the next years, so as appearance of new ones. According to threats prediction in [7], some of the threats that will become more aggressive and widespread in coming years are following: the rise of machine-to-machine attacks, propagation of worms in headless devices, and two-faced malware.

- **Machine-to-Machine Attacks** will take advantage of connected systems of mobile devices like connected medical devices and their host applications, connected home automation, smart TVs, and also connected home routers.
- **Worms in Headless Devices** refer to foreseen spread of worms within less complex devices, like smartwatches, by means of communication protocols.
- **Two-faced Malware** is type of malware designed to execute an innocent task to avoid detection system, and then, once it bypassed security checks, execute its malicious payload.

8.3 Malware Detection Solutions

With increased number of mobile threats the need to protect from them is growing, resulting in higher demand for effective detection systems. Reports indicate marked growth in the usage of anti-virus and anti-malware solutions for mobile platforms, which went from a 36 % rate of use in 2014 to 45 % in 2015 [1].

User's expectation from detection systems are that they are able to detect malware with high confidence without producing false positives and creating disturbance to regular usage. Additionally, any security mechanism targeted toward mobile systems should take their battery-operated characteristics into account as they may significantly limit the ability to run complex malware detection systems on devices. Providing detection mechanisms that are at the same time effective, able to detect variety of malware that exist today, and with low complexity, so that they do not significantly affect battery life, is a challenging task, and most of the proposed solutions are trade-off between these requirements.

Although number of threats is observed in variety of mobile devices, most of existing malware is targeting mobile phones and tablets. Due to this reason, most of current solutions are provided for them. In the rest of this section we discuss these solutions in more detail. Existing detection solutions can be divided in: signature-based, static, and dynamic.

8.3.1 Signature-Based Detection

A method that is commonly used in current anti-virus and anti-malware solutions is based on generation of representative signatures for existing malware samples and maintenance of a database consisting of them. Once the signature is recognised, malware is detected with high confidence. Although the number of false positives with such systems is low, they heavily rely on the maintenance of the database with signatures. Namely, it has to be frequently updated with new signatures that appear on the market. In mobile environment, this might be difficult due to the fact that the device is not constantly connected to the Internet, that sometimes is connected with mobile data that is charged, or that the device does not contain enough memory to store all available malware signatures.

8.3.2 Static Detection

These methods are focused on analysis of static features of applications (e.g. granted permissions, API calls, source code debagging) and discrimination between malware and trusted based on them.

One approach to static malware detection is proposed in [8] where high detection accuracy is achieved by using features from the manifest file and feature sets from disassembled code. Reported overhead is sub-linear. Its performance increases with $O(\sqrt{m})$, where m is the number of analysed bytes. Also the mechanism presented in [35] uses static features including permissions, Intent messages passing and API calls to detect malicious Android applications.

Apple, Google, and Nokia use application permissions and review to protect users from malware. The effectiveness of these mechanisms against malware in a given data set is evaluated in [16]. In [16], sending SMS messages without confirmation or accessing unique phone identifiers like the IMEI are identified as promising features for malware detection as legitimate applications ask for these permissions less often [17]. For example, nearly one third of applications request access to user location but far fewer request access to user location, and to launch at boot time. The authors concluded that although the number of permissions alone is not sufficient to identify malware, they could be used as part of a set of classification features, provided that all permissions common to the malware set are infrequent among non-malicious applications.

In [34], as a feature for detecting susceptibility of a device to malware infection, a set of identifiers representing the applications on a device is used. The assumption is that the set of applications used on a device may predict the likelihood of the device being classified as infected in the future. Nevertheless, observing just this feature is not enough to give precise answer about device being attacked due to low precision and recall [34].

In the nutshell, static detection is an effective approach in terms of resource consumption. However, due to the nature of this approach that analyses the applications only based on their static features, it is not able to detect malware that appears at run-time, it is prone to obfuscation [26], and cannot detect variations of existing malware samples that are easy to create and distribute.

8.3.3 Dynamic Detection

Dynamic detection appears as a promising candidate able to detect variety of malicious samples that currently exist on the market. The main advantage of this approach is that dynamic system features are observed at run-time, such as for example system calls and network behaviour, and based on them and previously trained models, detection is performed. In this way, by observing the behaviour of the system at run-time, systems are more resistant to variety of existing malware samples and more difficult to bypass. The reasoning behind is that while attackers can obfuscate the code itself it is difficult to obfuscate its behaviour.

Dynamic detection mechanisms are used in [10] to detect mobile worms, viruses and Trojans. The authors start with the extraction of representative signatures. Later on, a database with malicious patterns is created and Support Vector Machines are used in order to train a classifier with both trusted and malicious data. The evaluation of both emulated and real-world malware shows that dynamic detection not only results in high detection rates but also detects new malware which shares certain similarity with existing patterns in the database.

Power consumption, monitored through battery usage, also appears to be a promising approach [9]. One of the proposed solutions, VirusMeter [23], monitors and audits power consumption on mobile devices with a power model that accurately characterises power consumption of normal user behaviours. In [20] creation of a database with power signatures is proposed, where a new power signature collected while the system is used is compared with the ones already existing in the database. However, to what extent malware can be detected on phones by monitoring just the battery power remains an open research question [9].

SmartSiren, presented in [14], is a collaborative virus detection and alert system for smartphones. It performs statistical and abnormality monitoring, detects abnormalities at both device and network level, and in case alerts being detected issues alarm to the targeted population. This approach is tested and validated on viruses spreading via Bluetooth and SMS and Windows Mobile 5.0 Smartphone Edition. The used dataset consists of three weeks of SMS traces collected from Indian national cellular service provider. The reported overhead is 33.6 % of the total messages.

In [31] the approach to identify the most representative features to be observed on a phone running on Symbian operating system and then sent to the network for further investigation is presented. After receiving the information about these features on the server side it is decided if the phone state is abnormal or within expectations. Following five features are identified as informative and used: RAM

Free, User Inactivity, Process Count, CPU Usage, SMS Sent Count. More in detail, RAM Free indicates the amount of free RAM in kilobytes, User Inactivity tells if the user was active in the last ten seconds, Process Count indicates the number of currently running processes, CPU Usage represents the percentage of CPU usage, and SMS Sent Count represents the amount of SMS messages in the message directory. This approach is validated by using as a dataset simulation of normal behaviour of 10 frequently used applications at that time, and one malware sample.

In [36] a probabilistic approach on detection of malware propagating through Bluetooth and messaging services is presented. It observes unique behaviours of the mobile applications and the operating users on input and output constrained devices, and builds a Hidden Markov Model to learn application and user behaviours. Later, based on this knowledge, it identifies behavioural differences between malware and human users. The analysis is performed on Linux-based smartphone.

In [32], Andromaly, a framework for detecting malware, is proposed. It uses variety of features related to: touch screen, keyboard, scheduler, CPU load, messaging, power, memory, calls, operating system, network, hardware, binder, and LEDs, and compares False Positive Rate, True Positive Rate, and accuracy of the following detection algorithms: Bayes Net, Decision Tree J48, Histogram, K-means, Logistic Regression, and Naive Bayes. The algorithms that outperformed the others in detection of Android malware were Logistic Regression and Naive Bayes. The results were obtained using 40 trusted applications and four developed malicious samples, since no real malicious applications were available at that moment.

In [18] feature selection was performed on a set of run-time features related to network, SMS, CPU, power, process information, memory and Virtual memory. As a measure of features importance, Information Gain was used along with four classification algorithms: Naive Bayes, Random Forest, Support Vector Machines, and Logistic Regression. Results show that, in this scenario, Random Forest gives the best performance. Random Forest is a combination of different tree classifiers [11]. Although it is a powerful algorithm in achieving high accuracy of detection, it also has high complexity. Results have been obtained by considering 30 trusted and five malicious applications.

In [33] automatic way to detect malware by using combination of static and dynamic approach towards malware detection is presented. In order to extend coverage of dynamic detection, static detection is used as a first step, where the authors take into account applications Manifest file, decompiled code and requested permissions. Further, they analyse the application in sandbox tracking native API calls of the application taken into account. Malware samples taken into account are 136 000 applications from Asian and Google Play market and 7500 malicious samples. The system is accessible via web interface for all the users that would like to test the suspicious applications.

Work proposed in [12], is a crowdsourcing system that uses real traces of application behaviour collected from users. The traces are analysed in the network by usage of k-means clustering. Malware is detected by investigation of system calls, and the authors argue that the monitoring system calls are the most accurate way to detect malicious Android applications, since they provide detailed overview on the events. Dataset used is consisting of Trojan samples, more precisely, three samples of self-written malware and two real malware application samples.

In [15] Madam, a Multi-Level Anomaly Detector for Android Malware, is presented. Madam is a framework that detects intrusions and malware actions on Android devices. It does the detection by monitoring system OS events (system calls) and the user activity/idleness. The evaluation of the system is performed by using 10 real malware samples on Android Ice Cream Sandwich Version 4.1 Samsung Galaxy Nexus phone. The reported overhead of the approach is 3 % of memory consumption, 7 % of CPU overhead, and 5 % of battery. In order to use Madam, rooting of a phone is required.

In [13] another approach that also uses system calls is presented. The approach uses machine learning to learn connections between malicious behaviour (e.g. sending high premium rate SMS or cyphering data for ransom) and their execution traces and then exploit obtained knowledge to detect malware. As opposed to other systems, where a limited set of system calls is taken into account, in this work, all system calls are considered so as their sequences. The approach is tested on real device, with a dataset consisting of 20 000 execution traces and 2000 applications.

An approach presented in [25], takes into account dynamic features (memory and CPU) and their importance in malware detection. It analyses these features and their significance within the malware families they belong to, and takes into account the most indicative ones for each family. It concludes that some features appear as good candidates for malware detection in general, some features appear as good candidates for detection of specific malware families, and some others are simply irrelevant. For the analysis of importance of features, the authors use Principal Component Analysis.

A work proposed in [27] consists of two components: a host agent and a network service. The main purpose of the host agent is to acquire files and send them to the network service, whereas the network service performs analyses using multiple detection engines in parallel to determine whether a file is malicious or not.

Another proposed solution is Paranoid Android [28], which uses the anomaly detection principle. Based on phone execution traces, security checks are performed on the synchronised copy of the phone that runs on a server. The phone used in the analysis is HTC G1 phone, and on the server side QEMU was used. The results show that battery life is reduced about 30 % and CPU load about 15 %.

The drawback of dynamic detection methods is that such systems might be too complex for limited resources of mobile systems. In some cases, as previously mentioned, detection engines are offloaded to a cloud or a server, thus imposing new challenges to the system related to data transmission, communication overhead and data privacy. Additionally, although the systems based on dynamic detection are more resistant than the ones based on static, if the detection is based on observance of representative features, and the attacker develops completely new malware that does have different behavioural pattern from the learned ones, it might happen that the system would not be able to recognise it as malicious. Additional drawback is that while systems are trained only a limited number of execution paths can be taken into account, thus potentially not triggering the ones with malicious intent, which might be exposed only later during run-time execution of the application on a device.

8.4 Discussion

Increased number of mobile devices, together with their increased usage, attracted also the attacker's attention and motivated them to abuse these devices and get into possession of users credentials such as private and sensitive data. As a result of this, the increase in the number of encountered threats and their variety is observed. This trend, together with the most representative mobile threats is discussed and explained in more detail in Sect. 8.2.

Research community, anti-virus and anti-malware providers are trying to cope with the attacks and provide effective and efficient solutions for detection. The solutions have to be accurate in order not to disturb users with false alarms. Additionally, solutions have to be efficient, and thus suitable for limited resources of mobile devices. Existing detection methods, with their advantages and disadvantages, are discussed in Sect. 8.3. Additionally, in Table 8.1, tabular representation of state-of-the-art approaches is given, consisting of their characteristics related to: type of analysis, type of threats, detection technique, operating system, detection side, dataset, overhead, and publication year. Based on this information, we could spot different trends and challenges that we discuss in the following part of the section.

8.4.1 Type of Analysis

As discussed in Sect. 8.3, current existing detection methods can be dynamic and static. In some approaches both of these methods are combined. In Fig. 8.1 we present the distribution of existing detection methods with respect to type of used analysis: dynamic, static, and combined. As we can observe dynamic detection is prevailing. Due to previously mentioned weaknesses of static detection, reflected mainly in its inability to detect malicious behaviour at run-time, it is not surprising that the most research is focusing towards development of dynamic solutions. The main problem that a designer of a system may face when providing and developing dynamic solutions is the complexity, which might limit their applicability on constrained-resource devices.

8.4.2 Type of Threats

In Sect. 8.2 different existent threats that can affect mobile devices are discussed. Among them, as the most dominant and the widespread, malware is identified and different types of malware that currently exist are explained. In Fig. 8.2 we present the distribution of threats taken into account in existing detection methods. As we can see from Fig. 8.2 most of the existing solutions try to provide protection from malware in general, without particularly focusing on its subgroups. While this trend

Table 8.1 Tabular representation of state-of-the-art approaches

Type of analysis	Type of threat	Detection technique	Operating system	Detection side	Dataset	Overhead	Publication year	Reference
Static	Malware	Support vector machines	Android	Phone	123 453 apps, 5560 malware	Sub-linear with respect to the number of analysed bytes	2014	[8]
Static	Malware	kNN clustering	Android	Server	1500 trusted, 238 malware	Linear in the size of the problem	2012	[35]
Static	Malware	Matching with created database	Android, iOS	Server	103 695 apps	Bellow precision of hardware instrumentation	2013	[34]
Dynamic	Malware	Naive Bayes, logistic regression, random forest, support vector machines	Android	Phone	30 trusted 5 malware	Not reported	2013	[18]
Dynamic	Malware	Support vector machines	Android	Phone	1000 trusted, 1000 malware	Not reported	2015	[13]
Dynamic	Malware	Principal component analysis	Android	Phone	1080 malware	Not reported	2016	[25]
Dynamic	Mobile worms, viruses and Trojans	Support vector machines	Symbian	Phone	2 malware	Not reported	2008	[10]

(continued)

Table 8.1 (continued)

Type of analysis	Type of threat	Detection technique	Operating system	Detection side	Dataset	Overhead	Publication year	Reference
Dynamic	Viruses spreading via Bluetooth, SMS and MMS	Statistical and abnormality monitoring	Windows Mobile 5.0	Server	Three week SMS traces collected from a Indian national cellular service provider	33.6 % of total messages	2007	[14]
Dynamic	Zero day attacks, the attacks that can be detected with AV scanner	Virus scanner, dynamic taint analyses	Android HTC G1, QEMU	Server	Phone replicas	Battery life reduced 30 %, CPU load 15 %	2010	[28]
Dynamic	Malware	Anomaly detection	Symbian	Server	10 trusted, 1 malware	Not reported	2008	[31]
Combined	Viruses	Multiple detection engines	Nokia N800	Server	5 million signatures	Not reported	2008	[27]
Combined	Malware	Static, Dynamic	Android emulator	Server	136 000 trusted, 7500 malicious samples	Not reported	2013	[33]
Dynamic	Trojan horses	k-means	Android	Server	3 self-written malware apps, 2 real malware	Not reported	2011	[12]

Table 8.1 (continued)

Type of analysis	Type of threat	Detection technique	Operating system	Detection side	Dataset	Overhead	Publication year	Reference
Dynamic	Malware	Anomaly detection	Android Ice Cream Sandwich, Samsung Galaxy Nexus	Phone	10 malware	3 % of memory consumption, 7 % of CPU overhead, 5 % of battery	2012	[15]
Static	Malware	PART, Prism, nearest neighbour	Android	Collaborative	240 malware	Not reported	2009	[30]
Dynamic	Malware propagating via Bluetooth, MMS and SMS	Anomaly detection, hidden Markov models	Linux-based Smartphone	Phone	346 sms normal sequences, 27 abnormal sequences	Not reported	2010	[36]
Dynamic	Malware	Anomaly detection	Windows mobile	Phone or server	4 worms, 1 battery-depletion attack	Not reported	2008	[20]
Dynamic	Malware	Anomaly detection	Symbian	Nokia 5500	Power consumption data under normal operations and under media player and cashbook	1.5 % power consumption overhead	2009	[23]
Dynamic	Malware	Anomaly detection, logistic regression, Naive Bayes	Android	Device	20 trusted apps, 20 trusted tools, 4 developed malware apps	8.5 % of RAM, 5.52 % of CPU, 10 % of battery	2012	[32]

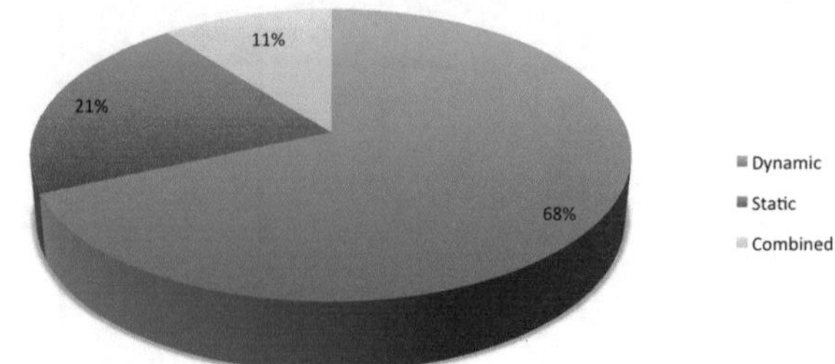

Fig. 8.1 Distribution of existing detection methods with respect to type of used analysis

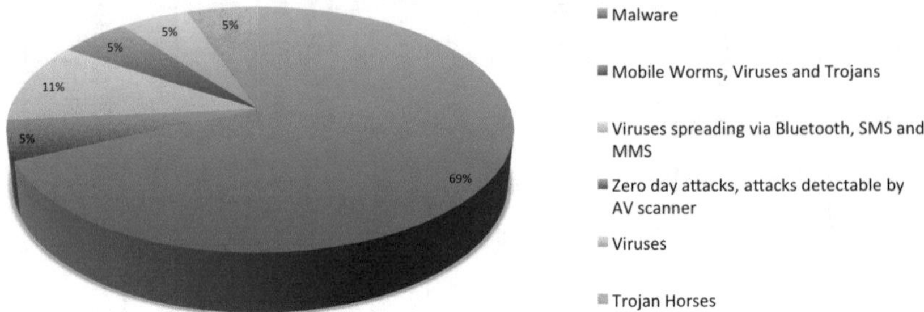

Fig. 8.2 Distribution of threats taken into account in existing detection methods

is understandable and makes used mobile devices more resistant to variety of threats, it also raises a question of the complexity of the solutions and its applicability on mobile devices. Namely, with the increased number of threats to protect from and their variety, also the solutions provided to detect them are increasing in complexity and computational overhead.

8.4.3 Detection Techniques

In order to detect malware, dynamic and static analysis can be used. In order to apply these approaches different detection techniques can be considered: anomaly detection, supervised and unsupervised learning, and clustering.

Anomaly detection is a technique particularly suitable when there are no many samples of malicious behaviour available. The main idea is to train the system with expected behaviour (normal or trusted), and then once something that goes out of ordinary happens, it is declared as an anomaly, outlier, or malicious activity. Mobile malware detection approaches based on the anomaly detection were particularly suitable in period when not so many real malicious samples were available.

Another way to detect malicious behaviour is to use unsupervised detection models. In this case, the developer is acquiring information about differences between

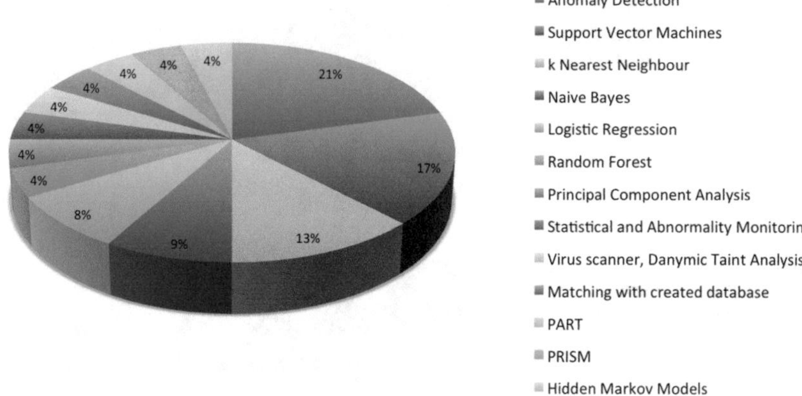

Fig. 8.3 Distribution of techniques taken into account in existing detection methods

trusted and malicious behaviour based on the information contained in data that can be reflected in different spacial structure, different values used, or different temporal characteristics. The most commonly used clustering method is kNN clustering, where the label of the data is assigned based on the labels of its nearest neighbours. kNN is a simple model, very suitable for limited resources of mobile systems.

Supervised learning is based on training, testing and development of models, where data is labelled as malicious or trusted. Once the model is trained and tested, it is deployed on a device. A variety of supervised learning models exist, both in terms of their complexity and detection performance. Depending on these parameters, models are more or less suitable for usage on mobile devices. Some of the supervised learning models, commonly being used in existing detection methods, are Support Vector Machines, Naive Bayes, Logistic Regression, and Random Forests.

In Fig. 8.3, distribution of detection techniques taken into account in existing detection methods is presented. As it can be observed, although the anomaly detection is slightly more used than the other techniques, there is no dominant technique and different detection algorithms can be suitable for different environments.

8.4.4 Operating System

In Fig. 8.4 distribution of operating systems taken into account in existing detection methods are illustrated. As we can see the most commonly used operating system is Android OS. This is due to its open structure and widespread usage that attracts both malware writers to abuse the systems and researchers to provide efficient and effective solutions to protect them.

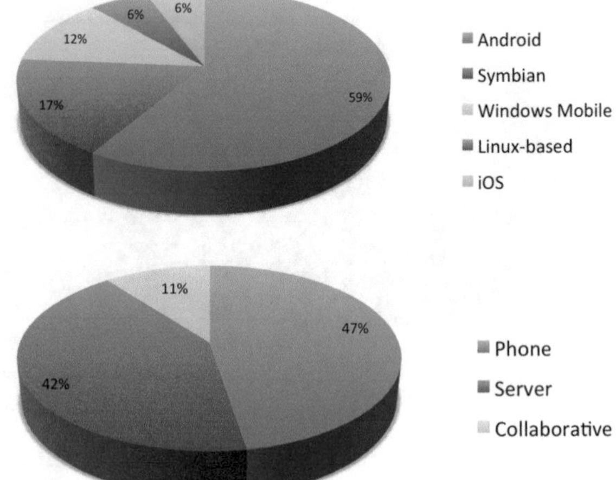

Fig. 8.4 Distribution of operating systems taken into account in existing detection methods

Fig. 8.5 Distribution of detection sides taken into account in existing detection methods

8.4.5 On Device Versus on Cloud Detection

Further separation of malware detection systems, as it is discussed in more detail in [24], can be done based on the detection side: on the device or on the cloud. Distribution of existing works with respect to the detection side is performed and obtained results are presented in Fig. 8.5. As we can see from Fig. 8.5 there is no dominant detection side, due to the fact that both approaches have their own advantages and disadvantages. The advantage of detection of malware on the device is that user data do not have to be sent into the network and, thus exposed to potential privacy breaches. Additionally, if the device is under attack a user receives an early notification about it, and so has more time to take appropriate countermeasures. However, computational capabilities of mobile devices sometimes limit the ability to run complex malware detection systems on them. Due to this reason, computations can be offloaded to the cloud where more sophisticated algorithms are used and detection is done with higher confidence. One of the drawbacks of this approach is that in case having no connection, user stays unprotected.

8.4.6 Datasets

Since the benchmark dataset for Android malware detection does not exist, researchers use different datasets to report their results. Fact that datasets are not always public is a significant limitation and makes comparison of research results difficult. It happens often that performance of some approaches are tested and reported only once on a specific dataset.

8.4.7 *Overhead*

One aspect of ultimate importance in mobile device detection, that was previously not a limitation for regular PC detection systems, is resource consumption. However, resource consumption is often missing in the evaluation of detection methods. This makes process of design of a suitable detection system difficult, since the designer cannot estimate in advance how complex the system is and whether scarce resources of devices are suitable for given applications.

8.5 Conclusions

With ever growing complexity of individual systems and of the entire network, rapidly increasing number of interconnected devices and continuously growing number of threats, designing secure and trustworthy mobile systems is a permanent challenge. In order to design such systems, both hardware and software threats have to be taken into account, and suitable solutions must be provided. In this chapter, we first give a short overview of the main hardware threats the system has to be protected from. Later, we describe and focus more on description of existing software threats, mostly reflected in malware, that affect mobile devices, as well as currently proposed solutions to cope with them. Additionally, the chapter discusses the main challenges and difficulties posed in front of designers during design process.

References

1. 2015 cyberthreat defense report. Tech. rep., CyberEdge Group (March 2015). http://www.brightcloud.com/pdf/cyberedge-2015-cdr-report.pdf.
2. 2016 threats prediction. Tech. rep., McAfee Labs (November 2015). http://www.mcafee.com/us/resources/reports/rp-threats-predictions-2016.pdf.
3. 2016 trend micro security predictions: The fine line. Tech. rep., Trend Micro (October 2015). http://www.trendmicro.com/cloud-content/us/pdfs/security-intelligence/reports/rpt-the-fine-line.pdf.
4. Internet security threat report volume 20. Tech. rep., Symantec (April 2015). https://www.symantec.com/content/en/us/enterprise/other_resources/21347933_GA_RPT-internet-security-threat-report-volume-20-2015.pdf.
5. It threat evolution in q2 2015. Tech. rep., Kaspersky Lab (July 2015). https://securelist.com/files/2015/08/IT_threat_evolution_Q2_2015_ENG.pdf.
6. Mcafee labs threats report. Tech. rep., McAfee Labs (August 2015). http://www.mcafee.com/us/resources/reports/rp-quarterly-threats-aug-2015.pdf.
7. New rules: The evolving threat landscape in 2016. Tech. rep., FortiGuard Labs (November 2015). http://www.fortinet.com/sites/default/files/whitepapers/WP-Fortinet-Threat-Predictions.pdf.
8. Arp D, Spreitzenbarth M, Hubner M, Gascon H, Rieck K. DREBIN: effective and explainable detection of android malware in your pocket. In: NDSS; 2014.

9. Becher M, Freiling FC, Hoffmann J, Holz T, Uellenbeck S, Wolf C. Mobile security catching up? Revealing the nuts and bolts of the security of mobile devices. In: Symposium on security and privacy. SP '11, IEEE Computer Society; 2011. p. 96–111.
10. Bose A, Hu X, Shin KG, Park T. Behavioral detection of malware on mobile handsets. In: 6th international conference on mobile systems, applications, and services (MobiSys). ACM; 2008. p. 225–38.
11. Breiman L. Random forests. Mach Learn. 2001;45(1):5–32. http://dx.doi.org/10.1023/A:1010933404324.
12. Burguera I, Zurutuza U, Nadjm-Tehrani S. Crowdroid: behavior-based malware detection system for android. In: Proceedings of the 1st ACM workshop on security and privacy in smartphones and mobile devices. SPSM '11. New York, NY, USA: ACM; 2011. p. 15–26. http://doi.acm.org/10.1145/2046614.2046619.
13. Canfora G, Medvet E, Mercaldo F, Visaggio CA. Detecting android malware using sequences of system calls. In: Proceedings of the 3rd international workshop on software development lifecycle for mobile. DeMobile 2015. New York, NY, USA: ACM; 2015. p. 13–20. http://doi.acm.org/10.1145/2804345.2804349.
14. Cheng J, Wong SH, Yang H, Lu S. Smartsiren: virus detection and alert for smartphones. In: 5th international conference on mobile systems, applications and services. MobiSys '07, ACM; 2007. p. 258–71.
15. Dini G, Martinelli F, Saracino A, Sgandurra D. Madam: a multi-level anomaly detector for android malware. In: Kotenko I, Skormin V, editors. Computer network security. Lecture notes in computer science, vol. 7531. Berlin Heidelberg: Springer; 2012. p. 240–53. http://dx.doi.org/10.1007/978-3-642-33704-8_21.
16. Felt AP, Finifter M, Chin E, Hanna S, Wagner D. A survey of mobile malware in the wild. In: 1st ACM workshop on security and privacy in smartphones and mobile devices (SPSM). ACM; 2011. p. 3–14.
17. Felt AP, Greenwood K, Wagner D. The effectiveness of application permissions. In: 2nd USENIX conference on web application development (WebApps). USENIX Association; 2011. p. 7.
18. Ham HS, Choi, MJ. Analysis of android malware detection performance using machine learning classifiers. In: 2013 international conference on ICT Convergence (ICTC). p. 490–5.
19. Kim CH, Quisquater J. Faults, injection methods, and fault attacks. IEEE Des Test Comput. 2007;24(6):544–5. http://doi.ieeecomputersociety.org/10.1109/MDT.2007.186.
20. Kim H, Smith J, Shin KG. Detecting energy-greedy anomalies and mobile malware variants. In: Proceedings of the 6th international conference on mobile systems, applications, and services. MobiSys '08. New York, NY, USA: ACM; 2008. p. 239–52. http://doi.acm.org/10.1145/1378600.1378627.
21. Kocher PC. Timing attacks on implementations of Diffie-Hellman, RSA, DSS, and other systems. In: Koblitz N, editor. CRYPTO '96, Proceedings of the 16th annual international cryptology conference on advances in cryptology, Santa Barbara, California, USA, August 18–22, 1996. Lecture notes in computer science, vol. 1109, Springer; 1996. p. 104–13. http://dx.doi.org/10.1007/3-540-68697-5_9.
22. Kocher PC, Jaffe J, Jun B. Differential power analysis. In: Wiener MJ, editor. CRYPTO '99, proceedings of the 19th annual international cryptology conference on advances in cryptology, Santa Barbara, California, USA, August 15–19, 1999. Lecture notes in computer science, vol. 1666. Springer; 1999. p. 388–97. http://dx.doi.org/10.1007/3-540-48405-1_25.
23. Liu L, Yan G, Zhang X, Chen S. Virusmeter: preventing your cellphone from spies. In: 12th international symposium on Recent Advances in Intrusion Detection (RAID). Springer; 2009. p. 244–64.
24. Milosevic J, Dittrich A, Ferrante A, Malek M. A resource-optimized approach to efficient early detection of mobile malware. In: 2014 ninth international conference on Availability, Reliability and Security (ARES). IEEE; 2014. p. 333–40.
25. Milosevic J, Ferrante A, Malek M. What does the memory say? Towards the most indicative features for efficient malware detection. In: CCNC 2016, The 13th annual IEEE consumer

communications and networking conference. Las Vegas, NV, USA: IEEE Communication Society; 2016.

26. Moser A, Kruegel C, Kirda E. Limits of static analysis for malware detection. In: Twenty-Third annual computer security applications conference, 2007. ACSAC; 2007. p. 421–30.
27. Oberheide J, Veeraraghavan K, Cooke E, Flinn J, Jahanian F. Virtualized in-cloud security services for mobile devices. In: 1st workshop on virtualization in mobile computing. MobiVirt '08, ACM; 2008. p. 31–5.
28. Portokalidis G, Homburg P, Anagnostakis K, Bos H. Paranoid android: versatile protection for smartphones. In: 26th Annual Computer Security Applications Conference (ACSAC). ACM; 2010. p. 347–56.
29. Quisquater J, Samyde D. Electromagnetic analysis (EMA): measures and counter-measures for smart cards. In: Attali I, Jensen TP, editors. Proceedings of the smart card programming and security, international conference on research in smart cards, E-smart 2001, Cannes, France, September 19–21, 2001. Lecture notes in computer science, vol. 2140. Springer; 2001. p. 200–10. http://dx.doi.org/10.1007/3-540-45418-7_17.
30. Schmidt AD, Bye R, Schmidt HG, Clausen J, Kiraz O, Yuksel K, Camtepe S, Albayrak S. Static analysis of executables for collaborative malware detection on android. In: IEEE international conference on communications, 2009. ICC '09; 2009. p. 1–5.
31. Schmidt AD, Peters F, Lamour F, Albayrak S. Monitoring smartphones for anomaly detection. In: 1st international conference on MOBILe wireless MiddleWARE, operating systems, and applications. MOBILWARE '08, ICST (Institute for Computer Sciences, Social-Informatics and Telecommunications Engineering); 2007. p. 40:1–40:6.
32. Shabtai A, Kanonov U, Elovici Y, Glezer C, Weiss Y. "Andromaly": a behavioral malware detection framework for android devices. J Intell Inf Syst. 2012;38(1):161–90. http://dx.doi.org/10.1007/s10844-010-0148-x.
33. Spreitzenbarth M, Freiling F, Echtler F, Schreck T, Hoffmann J. Mobile-sandbox: having a deeper look into android applications. In: Proceedings of the 28th annual ACM symposium on applied computing. SAC '13. New York, NY, USA: ACM; 2013. p. 1808–15. http://doi.acm.org/10.1145/2480362.2480701.
34. Truong HTT, Lagerspetz E, Nurmi P, Oliner AJ, Tarkoma S, Asokan N, Bhattacharya S. The company you keep: mobile malware infection rates and inexpensive risk indicators. CoRR abs/1312.3245; 2013.
35. Wu DJ, Mao CH, Wei TE, Lee HM, Wu KP. Droidmat: android malware detection through manifest and API calls tracing. In: 2012 seventh Asia joint conference on information security (Asia JCIS). p. 62–9.
36. Xie L, Zhang X, Seifert JP, Zhu S. PBMDS: A behavior-based malware detection system for cellphone devices. In: Proceedings of the third ACM conference on wireless network security. WiSec '10. New York, NY, USA: ACM; 2010. p. 37–48. http://doi.acm.org/10.1145/1741866.1741874.
37. Zhou Y, Jiang X. Dissecting android malware: characterization and Evolution. In: Proceedings of the 2012 IEEE symposium on security and privacy. SP '12. Washington, DC, USA: IEEE Computer Society; 2012. p. 95–109.

Chapter 9
Ring Oscillators and Hardware Trojan Detection

Paris Kitsos, Nicolas Sklavos and Artemios G. Voyiatzis

9.1 Introduction

The integrated circuit (IC) supply chain was considered as well protected for a long time now being constrained in one geographical location or even in one company. The trustworthiness of integrated circuits manufactured in remote silicon foundries with components sourced from parties spread around the world raises lately a lot of concern [1, 20, 27].

A hardware Trojan horse[1] is a modification of the original IC design by a malevolent actor aiming to exploit hardware characteristics or hardware mechanisms in order to access and manipulate information stored or processed on the chip. Trojans are not anymore a hypothesized threat but rather a realistic one.

In this chapter, we review the characteristics of hardware Trojans, taxonomies to classify their behavior and risks, and detection methods (Sect. 9.2). Then, we describe two novel Trojan works based on ring oscillator constructs. The first one focuses on true random number generators (TRNGs), commonly used in cryptographic hardware, and how a Trojan can interfere with their hardware implementation when ring oscillators are used as source of entropy. The second one explores a novel use of

[1]Herein, we will use the terms "Trojan" and "hardware Trojan" as synonyms for this term.

P. Kitsos (✉)
Computer and Informatics Engineering Department, TEI of Western Greece,
Antirio, Greece
e-mail: pkitsos@ieee.org; pkitsos@eap.gr

N. Sklavos
Computer Engineering and Informatics Department, University of Patras,
Patras, Greece
e-mail: nsklavos@ceid.upatras.gr

A.G. Voyiatzis
SBA Research, Vienna, Austria
e-mail: avoyiatzis@sba-research.org

N. Sklavos et al. (eds.), *Hardware Security and Trust*,
DOI 10.1007/978-3-319-44318-8_9

transient-effect ring oscillators (TEROs), as an on-chip mechanism to detect the contamination of a cryptographic algorithm with a Trojan (Sect. 9.3). Finally, we conclude by discussing an outlook for the future of hardware Trojans and detection techniques (Sect. 9.5).

9.2 Trojans and Trojan Detection Techniques

In this section, we describe the characteristics of hardware Trojan horses, taxonomies that have been proposed to classify them and design defenses. We also review selected defense methodologies that are proposed in the literature.

9.2.1 Trojan Characteristics

A first step toward estimating the risk of a hardware Trojan horse insertion and designing efficient and appropriate defenses is to consult a classification of Trojans. The engineering challenge is rather big, as malevolent actors can use a multitude of methods to modify the functionality of their target for their own benefit.

One classification is based on the characteristics of the Trojans [21]. As depicted in Fig. 9.1, the characteristics of Trojans are the type, the payload, the trigger, the size, and its communication protocol.

- The *type* of the Trojan is the kind of the attack to the pointed circuit. Based on that, the Trojan can be functional or parametric.
- The *payload* is the effect that the Trojan will have to the system, the gain that the attacker with have from that attack. For most of the Trojans, the primary point is to steal the secret information that the system is designed to securely send to the outside world.
- The *size* of the Trojan is another important factor. The area that can cover could be from a small wire to a large gate. Of course, it is more usual and efficient to the attackers to use small sized Trojans, while it will benefit them to make the exposure of the Trojan more difficult.
- The *communication protocol* is the way the Trojan will communicate and affect the circuit. It could be stored to a peripheral device or directly inside the chip.
- Last but not least is the *trigger* point of the Trojan. Trigger refers to the way the Trojan will be activated. Some of the most common activation ways include changes to the temperature or the voltages of the system. Another way is related to changes on the logic of the circuit. For example the data on an input, an unexpected interrupt or changes at the clock of the system. Of course there are systems that they do not have a trigger point, which means that the Trojan can be triggered at any time.

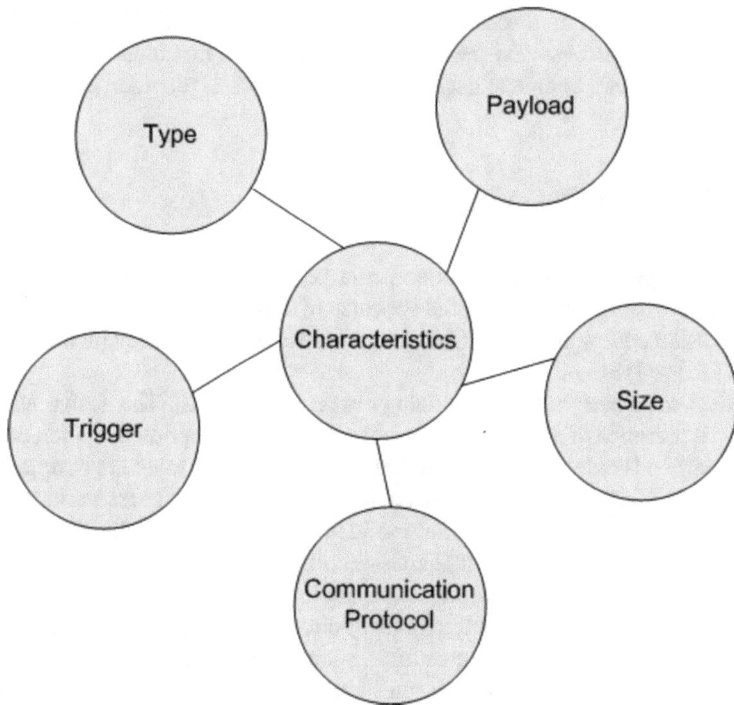

Fig. 9.1 Characteristics of hardware Trojans

9.2.2 *Trojan Taxonomies*

Many taxonomies for hardware Trojans are proposed in the literature, each shedding light from a different angle on the issue of coping with the effects of contaminated chips and malicious functionality [13, 18, 35]. Trust-HUB is an online resource for hardware security research and provides a rich taxonomy of Trojans too.[2]

An important part of the taxonomies is that they encompass information about the application scenario for a Trojan. Such information, beyond the Trojan characteristics themselves, are useful toward understanding the motivation of a malevolent actor, the possible stages of the supply chain where one might try to contaminate the hardware design, and the associated risks. These parameters are crucial into evaluating and quantifying the cost and the benefit (resp. damage) of an attack and thus, justify the resource investment toward designing and implementing appropriate technical and organizational countermeasures.

There is a general agreement in the available literature that the most concerning motivations for an attack based on hardware Trojans are the information leakage from the circuit and the denial-of-service against a device. In that sense, cryptographic

[2]https://www.trust-hub.org/resources/benchmarks.

primitives realized in hardware are an attractive target of a Trojan due the critical information they process and the enormous state space that must be searched for detecting it. A rather complete list of attack motivations is provided in [15].

9.2.3 Detection Techniques

There is a large number of works and a large variety of proposed methods and techniques focusing on detecting the presence of a hardware Trojan. Here, we summarize a selected few and the readers interested for a recent survey on the topic can consult [18] or [13].

A novel sustained vector methodology is presented in [2]. The design allows to expose the presence of a Trojan based on the power consumption deviation between the unmodified (golden) and the modified circuit. A careful selection of appropriate input vectors allows to magnify this deviation beyond the process variation level. The proposed technique further allows to identify the regions of the circuit that are more susceptible to contain the malicious circuitry. This methodology requires zero-overhead silicon. However, it requires access to the design of the circuit for the analysis phase and availability of trusted (golden) circuits to compare against (Fig. 9.2).

Measuring global, chip-wide quantities, such as the overall power supply current (I_{DDQ}) or the total execution time, can be a quite inefficient side-channel analysis approach. This is due to the low signal-to-noise ratios. More fine-grained measurements can provide better evidence, assuming that you can control the points of measurement on a chip. A region-based transient power signal analysis method leveraging statistical processing can reduce the impact of process variations and leakage currents and assist detection efforts [24]. A similar approach replaces transient power signal analysis with path delay information using shadow registers and a skewed shadow clock, once more leveraging statistical processing to reduce process

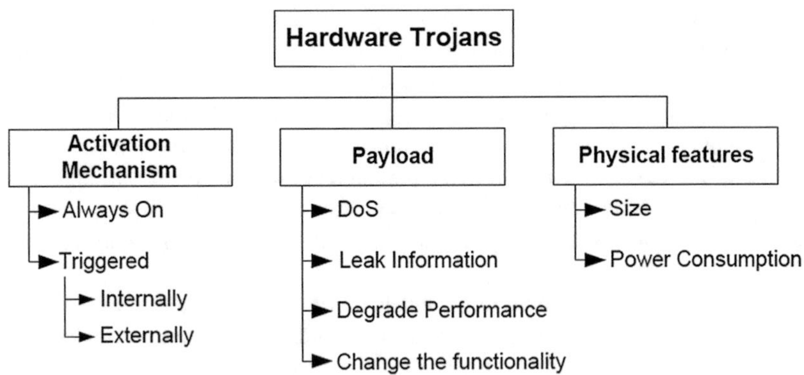

Fig. 9.2 A hardware Trojan horse taxonomy (*Source* [35])

variation [25]. However, it should be noted that the malicious actors can spread their Trojan in multiple regions and can avoid areas that are covered by the transient power analysis.

The use of dummy scan flip-flops is presented in [29]. This methodology adds extra (dummy) scan flip-flops in the circuit with the aim to increase the switching activity. The motivation is that with increased switching activity, the time to activate a Trojan that uses a rare signal transition will drop sufficiently enough. Thus, it becomes feasible to activate the hidden Trojan within the testing time envelope. The proposed methodology incurs a negligible area overhead for its implementation and supports a configurable threshold on the number (or percentage) of aimed transitions. We note that more advanced Trojans can use as an activation signal a rare *combination* of equiprobable transitions on the expense of more complex circuitry and thus, overcome the protection offered by this methodology [31]. Also, more advanced Trojans can use the scan signals themselves so as to hide its presence while on test mode [26]. This concept shares a lot of common ground with malicious software (malware) techniques used to detect that the software is executing inside an analysis environment and then adjusts its behavior accordingly [22, 34].

The voltage inversion technique to Ascertain Malicious Insertions in ICs (VITAMIN) is another approach to detect a malicious design embedded into an integrated circuit [3]. It utilizes the work in [2] and proposes an inverted voltage scheme, aiming to increase the frequency of gate transitions and thus, the activations of the Trojan. As it was noted earlier, this extended method also requires access to trusted circuits to compare against.

The secure heartbeat and dual-encryption (SHADE) architecture uses a different starting point [4]. SHADE assumes that the fabricated chips are already infected with a Trojan and builds a combined compiler-and-hardware system so as to provide a secure execution environment. This is achieved by preventing and by detecting a series of Trojans (mainly information leakage and DoS ones) in the expense of additional hardware modules for encrypting and decrypting on-the-fly memory contents and compiler instrumentation.

Hardware devices based on the USB interface can also be an attractive point for Trojan insertion, as they can infiltrate information from a computer on other network endpoint devices while leaving no traces in the network traffic. A proof-of-concept implementation of a fake USB device is described in [8, 9]. The associated risks and potential defenses are then discussed in [10]. The ability of self-reporting capabilities for USB devices is an inherent feature of the USB specification and thus, it is unclear how appropriate defenses can be devised without altering the specification.

The first in-silicon defense is reported in [14]. There, the attack model assumes that a hardware Trojan leaks information from an IC performing cryptographic operations. The information are leaked through a side-channel, namely the carrier modulation at the wireless interface of the IC. The proposed countermeasure monitors the transmission power and other parameters of the transmission signal; statistical processing of the collected information can reveal deviations that result in the exposure of the Trojan's presence.

The aforementioned Trojan attack scenarios and countermeasures indicate the need to design defenses that match the operating environment of a device and its envisioned function. In the next two sections, we focus on the use of ring oscillators as both a Trojan attack and defense vectors against secure hardware functions, namely, random number generation and stream encryption/decryption.

9.3 Trojan Detection in True Random Number Generators

The random number generators, either pseudo (PRNG) or true (TRNG), are an important component of modern security and cryptographic operations. This source of randomness is often used as a starting point for generating ephemeral or long-term cryptographic keys, for ensuring freshness of computed cryptographic tokens, and for protecting against replay attacks. As such, a TRNG can be an attractive target for hardware Trojan infection.

9.3.1 TRNG Design

TRNGs commonly use some kind of physical phenomenon as a source of entropy. Typically, these phenomena are analog. Therefore, an extraction mechanism is needed in order to convert the analog values into digital ones. Once the entropy source has been digitized, the statistical properties of the digitized signal will be evaluated with the purpose of establishing the TRNG quality. After this first evaluation, it is often conclude that a post-processing block is required to correct the output distribution. Finally, due to the importance of the TRNGs in security systems, it is recommendable to check the quality of the random output during its generation. Often, embedded (online) tests are employed to set an alarm when the generated output does not comply with some statistical requirements. Some well-known battery of tests used to assess the quality of a TRNG's output include DIEHARD,[3] ENT,[4] AIS31 [30], and NIST [28]. The typical blocks of an embedded TRNGs are depicted in Fig. 9.3.

9.3.2 Trojan Characteristics

The idea of embedding statistical tests as the last block of a TRNG design is rather interesting from a security point of view. This block acts as a guard that can detect deviations and manipulations of the input in the previous steps and raise an alert.

[3] http://stat.fsu.edu/pub/diehard.

[4] http://www.fourmilab.ch/random/.

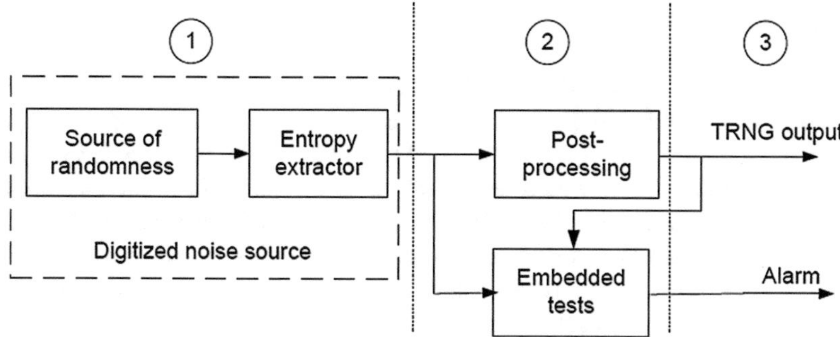

Fig. 9.3 General scheme of an embedded TRNG

A malevolent actor wishing to manipulate the operation of the TRNG must overcome this alert generation in order to launch a successful attack. There are four points of concern, which are analyzed in the following paragraphs.

9.3.2.1 Payload

We begin our analysis with the Trojan payload. This part performs the malicious action envisioned by the malevolent actor. As discussed in Sect. 9.2.2, the main lines of attacks are: denial-of-service (DoS), functionality changes, information leakage, and performance degradation. A DoS Trojan or a Trojan that introduces functionality changes should be easy to detect, once it is activated. This can be achieved by means of functional testing or even by the online tests that monitor TRNG's output quality.

The information leakage relates to the exfiltration of some secret information through a communication channel (side or not). In the case of a TRNG, this can relate to the transmission of a random number used afterwards as a cryptographic key or a nonce for a security protocol. We consider such kind of payloads beyond our scope. The TRNG cannot distinguish per se the envisioned use of its output; in order for an attacker to leak information by attacking the TRNG output, they would have to integrate such a large amount of semantic information so that it *should* be rather easy to detect the manipulation of the circuitry.

The performance degradation can be considered in two components. The first relates to the physical degradation caused by ageing effects. This can be achieved for example by supplying more current to the device or operating the device in higher temperature environments. Such kinds of physical rather than logical payloads are beyond our scope. The second component relates to the degradation of the TRNG output, lowering its statistical quality.

In the following, we will focus on Trojans, which aim to lower the statistical quality of the TRNG in a subtle way so that they remain undetectable by the online tests. We will use the notation of *T4RNG* instead of "Trojan-targeting-the-TRNG" for the sake of readability.

9.3.2.2 Activation

We safely assume that a T4RNG will not be always on. Rather, it will adjust its operation so as to attack the output of the TRNG at a specific point of time, where some critical information is to be generated (e.g., a new cryptographic key). Should the Trojan is always on, it would be easier to detect it during functional testing and through the embedded tests. Thus, T4RNG will be classified as trigger-based Trojans.

We consider both internally and externally triggered mechanisms in our study. In principle, a TRNG integrates some analog circuitry that interfaces with the environment and uses some physical quantities as entropy sources. Thus, it is possible to include external triggers so as to activate the Trojan, without injecting new components in the design that could be easily detected even by visual inspection (e.g., an extra antenna).

9.3.2.3 Physical Characteristics

The physical characteristics of a Trojan is an important consideration from an attacker point of view. The major concerns relate to the *size* and the *power consumption* of a Trojan. We discussed in the previous paragraph the issues of external triggers that require additional components. In more general terms, the overall size of the Trojan compared to the size of the original, unmodified IC is very crucial to avoid detection. There are TRNG designs with a really low number of logic gates (e.g., using only nine inverters and size registers) [5]. There, it is almost impossible to pass undetected. There are also many TRNG designs with many more logic gates [7]. In principle, one should not only consider the TRNG circuitry, but also the whole design of the TRNG, as depicted in Fig. 9.3.

Also, the power consumption of the malicious circuitry must be negligible compared to the overall power consumption. In the case of the TRNGs, this is not a great concern, as the TRNGs consume a lot of power for their operation. For instance, TRNGs often use jittery clocks as an entropy source. These clocks mainly comprise ring oscillators (ROs), self-timed rings (STRs), or phase-locked loops (PLLs). Such constructs are power hungry. Thus, the power consumption of an additional T4RNG would be rather small and indistinguishable in the process variation envelope of power traces.

9.3.2.4 Area of Injection

A T4RNG will be successful if it succeeds in lowering the quality of the TRNG output while at the same time it succeeds to pass the embedded statistical tests. Revisiting Fig. 9.3, we identify three points in space (zone 1, 2, and 3 respectively) for injecting the malicious functionality.

The effect of a Trojan that manipulates the entropy source (zone 1) will be probably canceled out or even detected by the embedded, online tests. For example, a simple and precise method to correlate the size of the jitter and the entropy of the generated but stream is presented in [12].

A T4RNG targeting the post-processing step (zone 2) will have an even harder task, as it needs to overcome two embedded tests, before and after the post-processing occurs. Also, it is possible that the post-processing algorithms nullify the Trojan's effect before the latter reaches the output.

There are two points that the T4RNG can target in the last block (zone 3). The first is to let the Trojan shut down (conditionally or not) the alarm signal. This should be easily detectable with conventional testing, as it resembles a "stuck-at" fault. The second option is to overcome the one last embedded test. As there is only one test to bypass, it might be possible to succeed. We explore the feasibility of this approach in the next section.

9.3.3 *Feasibility of a T4RNG*

We present a T4RNG design that fulfills the aforementioned characteristics. Our example consists of one XOR gate, one AND gate, and two ROs shifted in phase by 180°, as depicted in Fig. 9.4. The example draws from the frequency-injection attack against RO-based TRNGs of [23]. There, the authors can control the phase of several ROs placed in the same device by injecting a sine wave in the power supply. Here, we exploit the same mechanism as a trigger for our Trojan. While the Trojan is not triggered, the complementary outputs of the two ROs are XOR'ed together to produce a logical one. In this case, the TRNG output is not altered. When the Trojan is triggered, the two ROs are in phase and their XOR'ed output produces a logical zero. In this case, the TRNG output is discarded.

It is very important to affect only a few bits so as to pass the statistical tests. We evaluated this through simulations in the MathWorks Simulink environment. The ring oscillators were simulated using pulse-generator blocks. As a TRNG, we used a random-source block that passes already the statistical tests. Figure 9.5 depicts the unaltered bitstream generated by the random-source block. Figure 9.6 depicts the bitstream generated while multiple Trojan activations occurred. The effect of the Trojan is evident in the minor diagonal. However, this bias goes undetected by the statistical testing: the manipulated output still passes the online tests.

We realized an RO-based TRNG on a Xilinx Spartan-3E FPGA board. The TRNG incorporates 511 stages, as proposed in [7]. Our aim was to showcase the light-weightness of the example T4RNG. Figure 9.7 depicts the Trojan-free design, Fig. 9.8

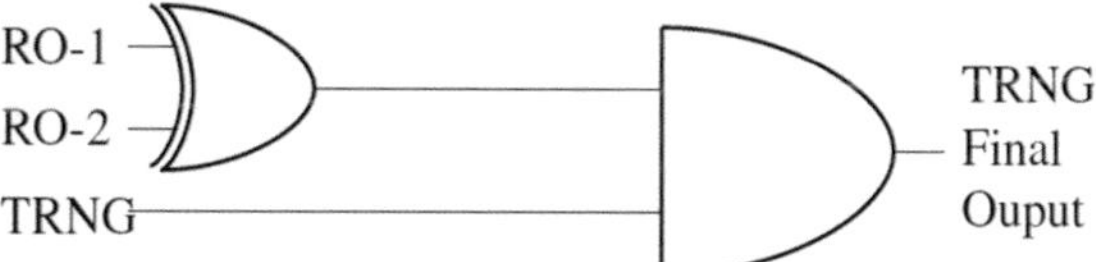

Fig. 9.4 Proposed Trojan

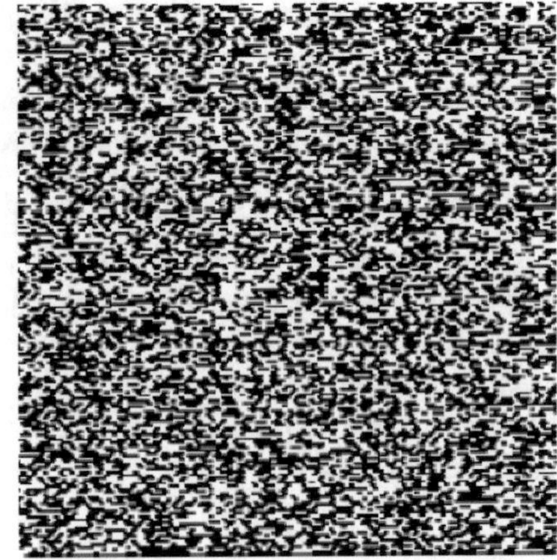

Fig. 9.5 Trojan-free TRNG output

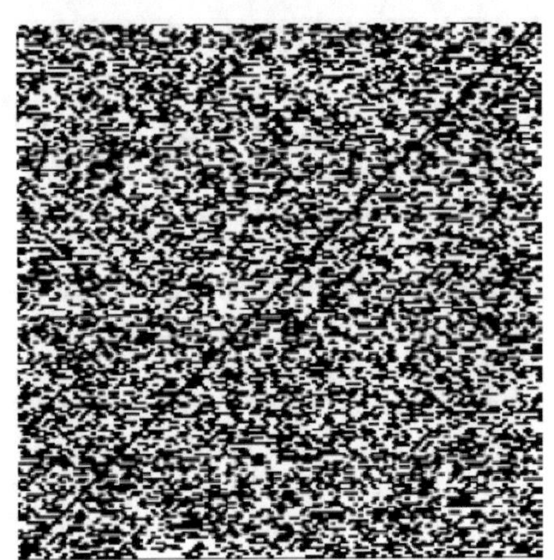

Fig. 9.6 Trojan-infected TRNG output

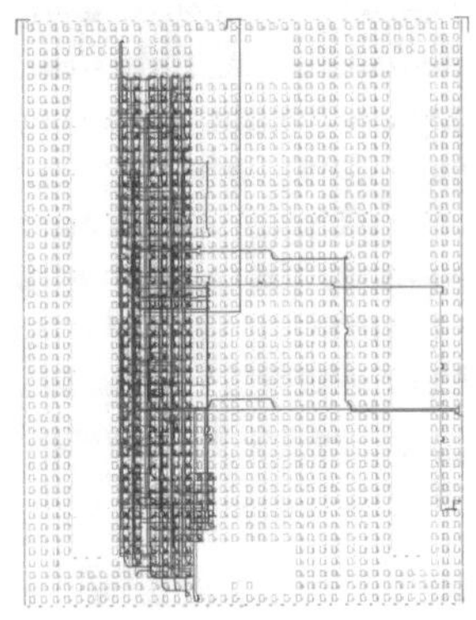

Fig. 9.7 Trojan-free TRNG

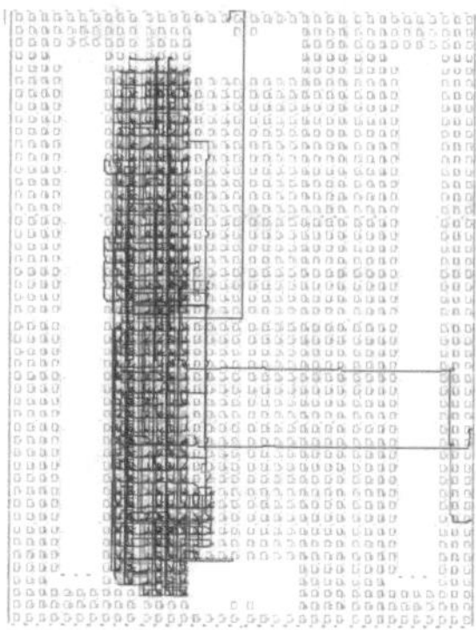

Fig. 9.8 Trojan-infected TRNG

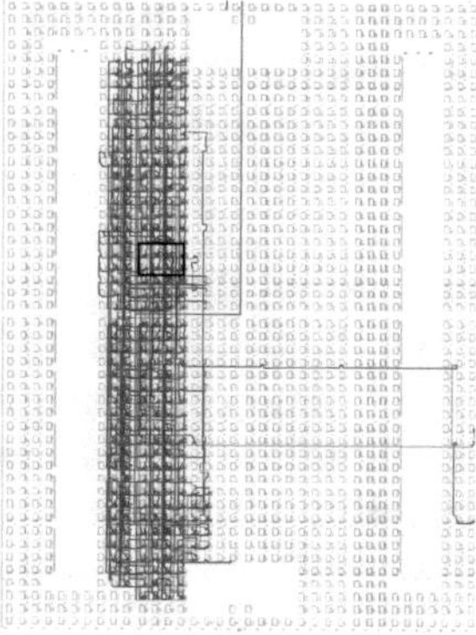

Fig. 9.9 Trojan part highlighted

depicts the design with the Trojan integrated, and Fig. 9.9 highlights the malicious part. It is evident that the malicious circuit consumes indeed a tiny space compared to the overall TRNG implementation.

In the best of our knowledge, this is the first report in the literature[5] that describes a hardware Trojan horse against a hardware implementation of a true random number generator. The Trojan design exploits the ring oscillators used as a source of randomness and succeeds in bypassing the embedded statistical testing albeit inserting a clearly identifiable bias in the output of the TRNG. It is evident that additional blocks for protection and detection must be incorporated in the TRNG hardware designs if used in critical environments such as for generating cryptographic key material. Toward this direction, we explore in the next sections the applicability of the ring oscillators which construct as a defense mechanism, for detecting the presence of hardware Trojans in a cryptographic-oriented hardware designs.

[5]An early version of this work was presented and discussed in TRUDEVICE 2015, a workshop collocated with DATE 2015, in Grenoble, France on March 2015.

9.4 Transient-Effect Ring Oscillators for Hardware Trojan Detection

The transient-effect ring oscillator (TERO) is a circuit that oscillates due to its inherent logic. The oscillation frequency depends on the exact components and the size of a circuit similarly to the case of a ring oscillator (RO). TERO was initially proposed for implementing a true random number generator (TRNG) [33]. Recently, it was proposed for implementing also a physically uncloneable function (PUF) [6]. In this section, we propose a novel use of TERO for hardware Trojan detection and we explore its applicability and efficiency as a extra tool for hardware Trojan horse detection. In the best of our knowledge, this is the first report in the literature[6] to explore the use of TERO for hardware Trojan detection.

A transient-effect ring oscillator (TERO) comprises an SR flip-flop implemented with two XOR gates and two AND gates [33]. This architecture has two control signals, for start and reset. The correct place-and-routing for a TERO is important so as to ensure the same length of the interconnections between the XOR gates.

Here, we use a simpler TERO architecture, where the XOR and AND gates are merged into NAND gates with some inverters in the feedback loop, as depicted in Fig. 9.10. The advantage of this approach is that only one control signal is used either for resetting or oscillating the TERO circuit. The reset occurs when the control signal, `ctrl`, is set to "0" and drives the loop to the same initial conditions before generating its output. When the control signal changes from "0" to "1", the TERO circuit starts to oscillate. An asynchronous counter (`counter`) is used in order to measure the TERO frequency.

9.4.1 Experimental Setup

In order to investigate the effectiveness of the TERO, we realized three hardware Trojan horses. The three Trojans target the SNOW3G stream cipher [32]. We briefly describe the SNOW3G algorithm and the Trojan designs used in our experiments.

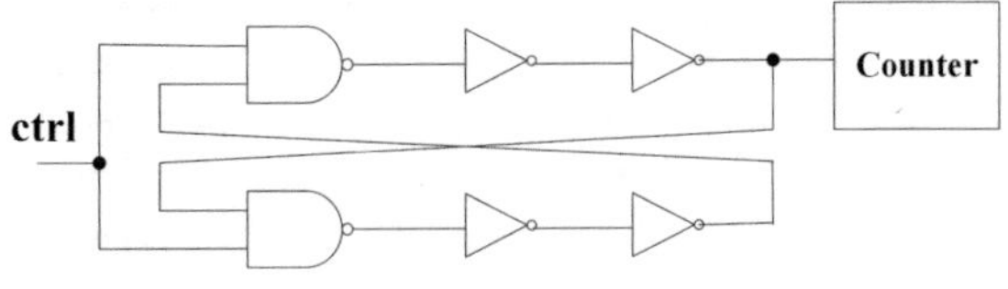

Fig. 9.10 TERO circuit

[6]An early version of this work was presented and discussed in TRUDEVICE 2015, a workshop collocated with DATE 2015, in Grenoble, France on March 2015.

9.4.1.1 The SNOW3G Stream Cipher

SNOW3G is a word-oriented stream cipher that generates a sequence of 32-bit words under the control of a 128-bit key and a 128-bit initialization variable. At first, a key initialization process is performed and the cipher is clocked without producing output. Then, the cipher operates in the key-generation mode and it produces a 32-bit ciphertext/plaintext word output in every clock cycle. The architecture of the SNOW3G cipher is depicted in Fig. 9.11.

9.4.1.2 Trojan Designs

The first one (T1), a combinational circuit, is formed as a tree of AND gates and the output of the tree is fed into a XOR gate that drives the system reset signal. T1 reads bits 24–31 of the output. If the bits are equal to `0xFF`, then T1 deactivates the reset signal.

T2 is a time bomb. It consists of a simple counter and an AND tree that reads bits 13–16 of the cipher's output. The tree output drives the enable signal of the Trojan counter. If the AND tree counts 100 sequences of "`111`" at the cipher's output, then T2 deactivates the SNOW3G reset signal.

T3 consists of two AND trees and two asynchronous counters. The first AND tree monitors input values at bits 13–16 of the cipher output and it is used so as to activate the first counter (`counter1`). If a sequence of "`1111`" occurs, then `counter1` is activated. Every second activation of `counter1`, T3 outputs and activation signal (`tmp_load`) that is used so as to trigger the second counter (`counter2`). The `tmp_load` signal is combined with three internal bits and, through the second AND tree, it is used to activate the second counter. After 62 pulses, `counter2` deactivates the cipher's reset signal.

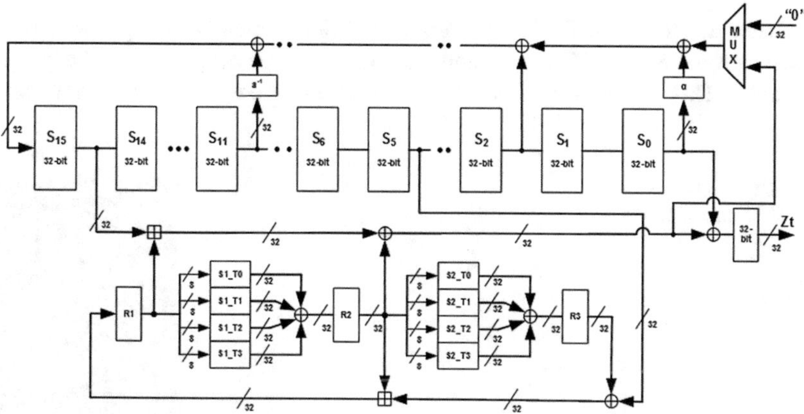

Fig. 9.11 SNOW3G hardware architecture

9.4.2 Experiments

A Xilinx Spartan 6 (XC6SLX75-2CSG484C) FPGA, the base of the SAKURA G board, was used in our experiments. The setup consists of the design of SNOW3G cipher enhanced with TERO and the circuits of the hardware Trojan horses. We used the implementation file extracted by the `PleanAhead` tool and especially the VHDL code of the Post-Place and Route simulation model. The model was simulated in order to derive the TERO oscillation frequency.

In our experiments, we used TEROs of different lengths so as to explore the optimal length for detecting the Trojan. The motivation for this comes from [17]. There, it is shown that the detection sensitivity is affected by the length of the RO used. The hardware Trojan horses occupied a small percentage of the available area and the TERO was placed in the circuit in a controllable fashion. In order to insert the hardware Trojan horses in a design implemented on FPGA, we use the hardware description language (HDL). While this method can be used to create many types of hardware Trojan horses, it is impossible to guarantee the exact place for the hardware Trojan horse insertion. If two systems are synthesized on the same FPGA board and they differ only to one hardware resource, the synthesis procedure will probably devise a completely different placing and routing.

In order to achieve efficient, fair, and most importantly, accurate measurements, one must build designs with the same place and route for a Trojan-free and for a Trojan-infected SNOW3G. This can be accomplished using BEL and LOC placement constraints. For the case of TERO, we have used parameterized area constraints. There are four designs: (a) a Trojan-free, containing SNOW3G and TERO, (b) the Trojan-free plus T1, (c) the Trojan-free plus T2, and (d) the Trojan-free plus T3. Snapshots of the four layouts are depicted in Figs. 9.12, 9.13, 9.14 and 9.15. In all cases, the TERO layout is shown as a white trace. It can be seen that the same layout for the circuits SNOW3G and TERO were created. This means that the hardware resources of the identical circuits are placed and routed on the same locations on the FPGA.

We decided to diffuse the Trojans around the cipher circuit and implement TERO in between the SNOW3G cipher and Trojans so as to better "sense" the process variations. This means that the greater distance between the counts means better

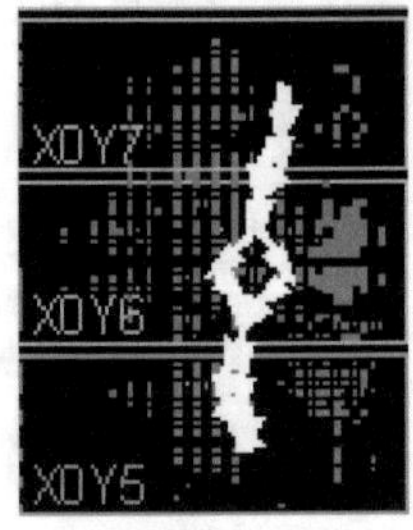

Fig. 9.12 Trojan-free SNOW3G and TERO

Fig. 9.13 SNOW3G, TERO, and T1

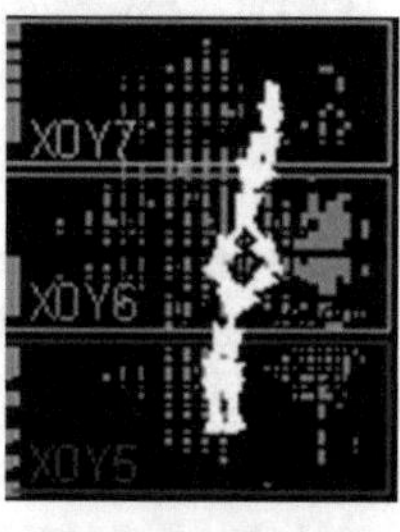

Fig. 9.14 SNOW3G, TERO, and T2

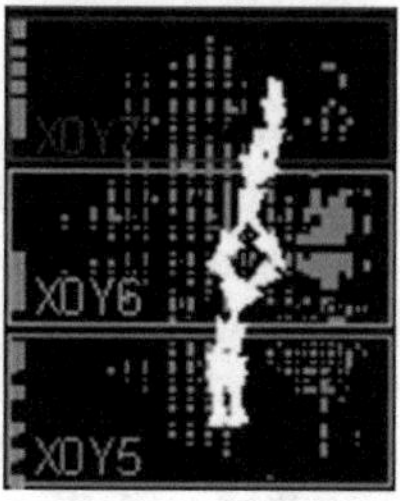

Fig. 9.15 SNOW3G, TERO, and T3

reliability and detection sensitivity of designs with hardware Trojan horses. The best metric for this is the absolute difference of oscillation counts between the Trojan-free and the infected circuits.

9.4.3 Results and Discussion

Table 9.1 summarizes the performance of TERO. It is clear that TERO is more sensitive when small lengths are used. Its sensitivity decreases as the lengths increase. In the case of T1, it cannot reliably detect the Trojan presence.

The results presented here are in accordance with the ones in [19]. In both cases, smaller lengths of TERO designs result in increased oscillation frequency and thus, detection sensitivity. Also, they confirm the result predicted by the theoretical work of [33]: the TERO designs oscillate at least the double frequency of equivalent RO designs. Thus, TERO designs with appropriate length are better than RO into

Table 9.1 Absolute differences for TERO counts

TERO length	TF1[a]	TF2[a]	TF3[a]
TERO 4	99	99	165
TERO 8	29	99	66
TERO 12	9	67	51
TERO 16	2	40	41
TERO 20	0	66	66
TERO 24	4	48	48

[a]TF*x*: Difference between Trojan-free and T*x*, respectively

detecting hardware Trojans. However, it will be beneficial to further explore in the future the applicability of this result using more complex Trojans and testing against other cryptographic algorithms and security constructs.

9.5 Conclusions and Outlook to the Future

Malicious hardware is not anymore a hypothesized but rather a realistic threat. It is possible to inject a hardware Trojan horse in one of the many stages of the hardware design and fabrication flow. Malevolent actors become more and more sophisticated and improve their capabilities. Applying a process of secure development and rigorous testing at each production stage, such as the one proposed in [11], can reveal circuit manipulations early enough so as avoid disastrous effects after fabrication. Yet, the possibility of receiving a chip with hidden Trojan functionality cannot be excluded.

Detecting the presence or the operation of a hardware Trojan horse requires an arsenal of tools, methods, and techniques. Each of them may be able to conclude that a given parameter or characteristic of an IC under test is within the acceptable limits of operation. Yet, even then, it is not sufficient so as to ensure the absence of a hardware Trojan inside the tested IC that operates under the radar or that is activated under a specific pattern. This discloses nothing more for the rest of the chips that are on the same production batch but was not feasible to test due to budget and time constraints.

From an attacker point of view, the two major obstacles to overcome for inserting a Trojan relate to the *moment* of the insertion during the design and fabrication workflow and to the *controllability* of the Trojan activation while remaining hidden during the various tests of the circuits.

The required infrastructure for testing and the increasing sophistication of the techniques indicate that it may not be possible to collect all resources under the same roof. Rather, a collaborative approach, such as the one pursued in the context of the TRUDEVICE network (http://www.trudevice.com/) may be preferable, where research teams and institutions with different skills and equipment combine forces as to achieve economies of scale, reproducibility of the experiments, and detection techniques using different equipment.

In this context, we presented two novel hardware Trojan works that target cryptographic constructs implemented in hardware. The focus of our contribution is on the use of ring oscillators as an attack and as a defense vector. The first relates to the manipulation of the ring oscillator output when used as a source of entropy for implementing in hardware a true random number generator (TRNG). We showed that it is possible to bypass embedded online testing of conformance and produce output patterns with detectable patterns. Such patterns can be a stepping stone toward launching more complex attacks on cryptographic algorithms. Even the slightest knowledge of key bits produced by a TRNG can be beneficial for an attacker.

The second relates to the use of ring oscillators for detecting the presence of a hardware Trojan. We showed that transient-effect ring oscillators (TERO) can have further uses than those proposed already in the literature as constructs of TRNGs or PUFs: TERO can also be used to detect the presence of a hardware Trojan. Furthermore, we showed that the length of a TERO affects its oscillation frequency and that shorter TEROs exhibit higher frequencies thus, they are more sensitive compared to conventional ROs to the presence of Trojans.

Drawing the experience of the software world, malware is still an unsolved problem after 40 years of existence. It would be unrealistic to expect a solution for the hardware malware immediately. However, we should aim for appropriate proactive and reactive defenses, as well as testing methodologies and practices, such as the one proposed in [16]. These can reduce the risk of hardware Trojan injection in first place and increase our trust on these devices that they operate as specified, without hidden functionality that can harm any part of our society.

Acknowledgements This work was supported in part by the EU COST Action IC1204 Trustworthy Manufacturing and Utilization of Secure Devices (TRUDEVICE), the GSRT Action "KRIPIS" with national (Greece) and EU funds, in the context of the research project "ISRTDI" while P. Kitsos and A.G. Voyiatzis were with the Industrial Systems Institute of the "Athena" Research and Innovation Center in ICT and Knowledge Technologies, and the COMET K1 program by the Austrian Research Promotion Agency (FFG), while A.G. Voyiatzis was with SBA Research.

References

1. Adee S. The hunt for the kill switch. IEEE Spectrum. 2008;45(5):34–9.
2. Banga M, Hsiao MS. A novel sustained vector technique for the detection of hardware Trojans. In: 2009 22nd international conference on VLSI design. IEEE; 2009. p. 327–32.
3. Banga M, Hsiao MS. VITAMIN: voltage inversion technique to ascertain malicious insertions in ICs. In: HOST'09, IEEE international workshop on hardware-oriented security and trust, 2009. IEEE; 2009. p. 104–7.
4. Bloom G, Narahari B, Simha R, Zambreno J. Providing secure execution environments with a last line of defense against Trojan circuit attacks. Comput Secur. 2009;28(7):660–9.
5. Böhl E, Ihle M. A fault attack robust TRNG. In: 2012 IEEE 18th international on-line testing symposium (IOLTS). IEEE; 2012. p. 114–7.

6. Bossuet L, Ngo XT, Cherif Z, Fischer V. A PUF based on a transient effect ring oscillator and insensitive to locking phenomenon. IEEE Trans Emer Top Comput. 2014;2(1):30–6.
7. Cherkaoui A, Fischer V, Fesquet L, Aubert A: A very high speed true random number generator with entropy assessment. In: Cryptographic hardware and embedded systems-CHES 2013. Springer; 2013. p. 179–96.
8. Clark J, Leblanc S, Knight S. Hardware Trojan horse device based on unintended USB channels. In: NSS'09, third international conference on network and system security, 2009. IEEE; 2009. p. 1–8.
9. Clark J, Leblanc S, Knight S. Compromise through USB-based hardware Trojan horse device. Future Gener Comput Syst. 2011;27(5):555–63.
10. Clark J, Leblanc S, Knight S. Risks associated with USB hardware Trojan devices used by insiders. In: 2011 IEEE international systems conference (SysCon). IEEE; 2011. p. 201–8.
11. Dabrowski A, Hobel H, Ullrich J, Krombholz K, Weippl E: Towards a hardware Trojan detection cycle. In: 2014 ninth international conference on availability, reliability and security (ARES); 2014. p. 287–94.
12. Fischer V, Lubicz D. Embedded evaluation of randomness in oscillator based elementary TRNG. In: Cryptographic hardware and embedded systems-CHES 2014. Springer; 2014. p. 527–43.
13. Jin Y. Introduction to hardware security. Electronics. 2015;4(4):763–84.
14. Jin Y, Makris Y. Hardware Trojans in wireless cryptographic ICs. IEEE Des Test Comput. 2010;27(1):26–35.
15. King ST, Tucek J, Cozzie A, Grier C, Jiang W, Zhou Y. Designing and implementing malicious hardware. LEET. 2008;8:1–8.
16. Kitsos P, Simos D, Torres-Jimenez J, Voyiatzis A. Exciting FPGA cryptographic Trojans using combinatorial testing. In: 26th IEEE international symposium on software reliability engineering (ISSRE 2015), IEEE CPS (2015). Gaithersburg, MD, USA, November 2–5, 2015. p. 69–76.
17. Kitsos P, Voyiatzis A. FPGA Trojan detection using length-optimized ring oscillators. In: 17th EUROMICRO conference on digital system design (DSD 2014). Verona, Italy: IEEE CPS; 2014.
18. Kitsos P, Voyiatzis A. Towards a hardware Trojan detection methodology. In: 2nd EUROMICRO/IEEE workshop on embedded and cyber-physical systems (ECYPS 2014). Budva, Montenegro; 2014.
19. Kitsos P, Voyiatzis A. A comparison of TERO and RO timing sensitivity for hardware Trojan detection applications. In: 18th EUROMICRO conference on digital system design (DSD 2015). Madeira, Portugal: IEEE CPS; 2015.
20. Lee W, Rotoloni B: Emerging cyber threats report 2013. Georgia Tech Cyber Secur Summit. 2012.
21. Lin L, Kasper M, Güneysu T, Paar C, Burleson W. Trojan side-channels: lightweight hardware Trojans through side-channel engineering. In: Cryptographic hardware and embedded systems-CHES 2009. Springer; 2009. p. 382–95.
22. Lindorfer M, Kolbitsch C, Comparetti PM. Detecting environment-sensitive malware. In: Recent advances in intrusion detection. Springer; 2011. p. 338–57.
23. Markettos AT, Moore SW. The frequency injection attack on ring-oscillator-based true random number generators. In: Cryptographic hardware and embedded systems-CHES 2009. Springer; 2009. p. 317–31.
24. Rad RM, Wang X, Tehranipoor M, Plusquellic J. Power supply signal calibration techniques for improving detection resolution to hardware Trojans. In: Proceedings of the 2008 IEEE/ACM international conference on computer-aided design. IEEE Press; 2008. p. 632–9.
25. Rai D, Lach J. Performance of delay-based Trojan detection techniques under parameter variations. In: HOST'09, IEEE international workshop on hardware-oriented security and trust, 2009. IEEE; 2009. p. 58–65.
26. Ray S, Yang J, Basak A, Bhunia S. Correctness and security at odds: post-silicon validation of modern SoC designs. In: Proceedings of the 52nd annual design automation conference, DAC '15. New York, NY, USA: ACM; 2015. p. 146:1–146:6.

27. Rogers M, Ruppersberger CD. Investigative report on the US national security issues posed by Chinese telecommunications companies Huawei and ZTE: a report. US house of representatives; 2012.
28. Rukhin A, Soto J, Nechvatal J, Smid M, Barker E. A statistical test suite for random and pseudorandom number generators for cryptographic applications. DTIC document: Tech. rep; 2001.
29. Salmani H, Tehranipoor M, Plusquellic J. A novel technique for improving hardware Trojan detection and reducing Trojan activation time. IEEE Trans Very Large Scale Integr VLSI Syst. 2012;20(1):112–25.
30. Schindler W, Killmann W. Evaluation criteria for true (physical) random number generators used in cryptographic applications. In: Cryptographic hardware and embedded systems-CHES 2002. Springer; 2003. p. 431–49.
31. Sreedhar A, Kundu S, Koren I. On reliability Trojan injection and detection. J Low Power Electron. 2012;8(5):674–83.
32. UEA2&UIA I. Specification of the 3GPP confidentiality and integrity algorithms UEA2& UIA2. Document 2: SNOW 3G specifications. Version: 1.1. ETSI; 2006.
33. Varchola M, Drutarovsky M. New high entropy element for FPGA based true random number generators. In: Cryptographic hardware and embedded systems, CHES 2010. Springer; 2010. p. 351–65.
34. Vidas T, Christin N. Evading android runtime analysis via sandbox detection. In: Proceedings of the 9th ACM symposium on information, computer and communications security. ACM; 2014. p. 447–58.
35. Wang X, Tehranipoor M, Plusquellic J. Detecting malicious inclusions in secure hardware: challenges and solutions. In: HOST 2008, IEEE international workshop on hardware-oriented security and trust, 2008. IEEE; 2008. p. 15–9.

Chapter 10
Notions on Silicon Physically Unclonable Functions

Mario Barbareschi

10.1 Introduction

The opportunity of extracting physical characteristics from fabric-induced variability for integrated circuits (ICs) has been representing the most important breakthrough for the semiconductor security. Like human fingerprints, retina blood vessels and DNA, which are the main identification means in the biometrics field, physically imprinted random patterns were adopted for the identification of serial manufacturing products [36, 42]. Such a technological innovation led to formally define the Physical Random Function [14], and then the physical(ly) unclonable function, or PUF, which are able to identify a silicon device by means of random manufacturing variability.

During last years, PUF has begun a hot topic in the field of hardware security and trust, in fact countless PUF circuits and architectures have been proposed in the literature. Actually, more and more proposed secure infrastructures and applications rely on the adoption of PUFs, specially because they guarantee extremely attractive properties, such as uniqueness, unclonability, anti-tamper, and intrinsically randomness, without requiring any modification to the classical photolithography manufacturing process.

The goal of this chapter is to collect most of concepts related to PUFs which have been divulged in scientific papers, trying to give a uniform view of formal notions and quality metrics. Furthermore, some pointers regarding the security and reliability issues are summed up, covering indispensable aspects to implement any PUF-based application

This chapter is structured as follows. Section 10.2 gives a formal characterization of PUFs and related terminology; and lists PUF properties, mostly covering what has been introduced in the literature. Moreover, it contains the related terminology, necessary to comprehend further notions and concepts. Section 10.3 concertizes

M. Barbareschi (✉)
DIETI—Department of Electrical Engineering and Information
Technologies, University of Naples Federico II, Via Claudio, 21 - 80125 Naples, Italy
e-mail: mario.barbareschi@unina.it

N. Sklavos et al. (eds.), *Hardware Security and Trust*,
DOI 10.1007/978-3-319-44318-8_10

previously defined properties by means of quality parameters (such as uniqueness, Sect. 10.3.1, and reliability, Sect. 10.3.2), which are used to compare and analyze available PUF architectures. Section 10.4 classifies existing PUF architectures by means of the source of randomness that their exploit.

10.2 A Formal Perspective on PUF

This section introduces the definition of a PUF and the properties which characterize almost every PUF implementation [14, 29]. Let a PUF be a mathematical application which associates inputs to outputs. Ideally such application is a function:

$$\theta \in \Theta : C \rightarrow R | \theta(c) = r, \ c \in C, \ r \in R. \tag{10.1}$$

The domain of $\theta(\cdot)$, the C set, is defined as the allowable inputs set, normally defined as challenges set, while the codomain, the R set, is called responses set. Θ includes all the produced PUF instances. The definition of $\theta(\cdot)$ does not contain other properties of the input–output mapping as, from the end-user point of view, any detail or physical parameter is hidden (cannot be *known*) and generated by manufacturing variability (cannot be *controlled*). Indeed, the mapping process which characterizes $\theta(\cdot)$ is constrained by parameters of gates and interconnections that determines the physical value exploited by the circuit. This means that each response is strictly correlated with a challenge trough physical quantities.

The pair $(c, \theta(c)) = (c, r)$ is defined as challenge/response pair (CRP). The PUF can be considered as a circuit which is able to provide a CRPs set which is unique for each device.

Being electrical, the relation among challenges and responses might turn out to be not a function in the mathematical sense: thermal noise, voltage fluctuations, temperature variations, and so on, making the physical parameters time variant. Such variability can be taken into account by substituting the response r with a random variable χ_r in definition (10.1). The distribution of χ depends on the manufacturing and implementation technology, on the PUF architecture and on the considered environmental conditions. From the probability point of view, r is the most probable response that can be generated from the PUF by stimulating it with the same challenge c, hence the challenge–response relation is substituted by $\theta(c) \approx r$. Higher is the probability to have $\chi_r = r$ as response from c, more reproducible, i.e., stable, is the PUF.

10.2.1 Unclonability

The main property of a PUF is the unclonability, that can be formally described as the impossibility to obtain a PUF θ' which is identical to θ, such that the latter can replace the former. Being a physical object, the unclonability can be considered in a mathematical or physical manner. Mathematical unclonability implies that it is hard to find a mathematical procedure that is able to provide the same CRPs set of θ (10.2a), while the physical unclonability indicates that it is hard to reproduce a device with a PUF θ' which can be recognized as θ (property 10.2b).

$$\nexists f : \theta(c) = f(c) = r, \ \forall c \in C. \tag{10.2a}$$

$$\nexists \theta' : \theta(c) = \theta'(c) = r, \ \forall c \in C. \tag{10.2b}$$

How these properties are guaranteed is determined by the manufacturing process and by the internal PUF design. The mathematical unclonability is guaranteed by the non-feasibility of modeling the manufacturing variability process and its impact on each physical parameter which determines the PUF behavior. The impossibility of producing two identical devices, even with the best knowledge on physical parameters affected by variability, is given by the noncontrollability of the manufacturing variability.

10.2.2 Uniqueness

The PUF uniqueness property is formulated as:

$$\nexists \widetilde{\theta} \in \Theta : \widetilde{\theta}(c) = \theta(c) = r, \ \forall c \in \widehat{C} \subseteq C. \tag{10.3}$$

The definition could be interpreted in an ambiguous way, since it seems to be overlapped, or at least very similar, to unclonability statements (Eqs. 10.2a and 10.2b). However this is not true, because the requirement to be unique for a θ is the nonexistence of another θ' that specifically belongs to Θ, unlike the unclonability that requires the nonexistence of a generic function that is able to substitute θ. Moreover, the uniqueness requires to be valid only for a subset of C.

Apart from the theoretical meaning, the uniqueness has a practical usefulness. Indeed, let $\dot{c} \in C$ be a random picked value and $\dot{r} = \theta(\dot{c})$. The pair $(\dot{c}, \theta(\dot{c}))$ induces a partition into Θ such that $\dot{\Theta} \cup \bar{\Theta} = \Theta$ and $\dot{\theta}(\dot{c}) = \dot{r}, \ \forall \dot{\theta} \in \dot{\Theta}$ and consequently $\bar{\theta}(\dot{c}) \neq \dot{r}, \ \forall \bar{\theta} \in \bar{\Theta}$. A successive picking of $\ddot{c} \in C$ defines another pair $(\ddot{c}, \theta(\ddot{c}))$ which causes a partition in $\dot{\Theta}$, and so on. By nesting this approach, so successive applications of other c values on θ, the picked CRPs set $\Psi = \{(\dot{c}, \theta(\dot{c}))\}$ will cause increasingly smaller partitions, to have a singleton set in which only one function, θ, is characterized by such CRPs. The success of this procedure and the required steps to complete it depends on the Θ cardinality, the characteristics of the PUF θ and on the picked c.

10.2.3 Unpredictability

The unpredictability property can be directly inherited from the definition of mathematical unclonability, Eq. 10.2a.

$$\Psi = \{(c, \theta(c))\}, \ \nexists \Phi : \Phi\left(\Psi, c_p\right) = \theta\left(c_p\right) = r_p, \ \forall c_p \in C. \tag{10.4}$$

What is requested to be unpredictable for a function is the inability to create a procedure Φ that, having a certain amount of challenge-response pairs Ψ for a PUF θ, is able to provide the same output of θ for a generic challenge c. The existence of this procedure is in direct contrast with the unclonability because Φ represents a mathematical clone that can predict the θ responses. Moreover, in the degeneracy case when $\Psi = \varnothing$, the statement 10.4 is equivalent to the Eq. 10.2a

10.2.4 One-Way Property

Formally θ is a one-way function $\iff$ given $r = \theta(c)$ it is hard to find $\lambda : \lambda(r) = \bar{c}$ and $\theta(\bar{c}) = r, \forall \bar{c} \in \bar{C} \subseteq C$. As for hash functions, in this definition "hard" is meant in the computational theory sense, so that given one output r of a PUF θ, it is very expansive, in terms of computation resources and time, to find one input $\bar{c}$ such that $\theta(\bar{c}) = r$.

10.2.5 Feasibility

Being an integrated circuit, a PUF inevitably introduces an overhead in area and time. As for the occupied area, the circuit has to extract the physical information and maybe implementing the challenge/response mechanism. As for the time overhead, the response extraction could require a significant amount of time, especially when the architecture is provided with a post-processing algorithm that has to be ran. Given $\theta \in \Theta$ and $c \in C$, θ is feasible if it is not hard to evaluate $\theta(c)$.

10.2.6 Tamper-Evident

The PUF θ is tamper-evident if any attempt to tamper the circuit permanently changes its CRPs set, obtaining a new PUF $\theta' : \theta'(c) \not\approx \theta(c), \forall c \in C$.

10.3 Quality Measurement on Silicon PUFs

In the literature, many silicon PUF architectures have been introduced and each one tries to enhance one of the properties illustrated in the Sect. 10.2. To compare them with objective measurements, some metrics have been introduced [31, 47]. The need for fair metrics to compare the quality of PUF proposals has generated some quality parameters.

10.3.1 Uniqueness

Ideally, due to the manufacturing variability, devices are unique in terms of physical quantities which characterize them. Silicon PUFs discretize such physical quantity to extract bit strings as responses from provided challenges, hence the uniqueness can be estimated trough responses comparison. Having a pair of PUF instances which provide N-bit responses, respectively $r_i = (r_{i,0},\ r_{i,1},\ \dots\ r_{i,N-1})$ and $r_j = (r_{j,0},\ r_{j,1},\ \dots\ r_{j,N-1})$, the uniqueness can be estimated as the fractional Hamming distance between r_i and r_j. The Hamming distance (HD) is a function computed over two binary strings of equal length which returns the number of homologous bits that differ, or, in other words, the minimum amount of substitutions needed to change one string into the other. The fractional HD (fHD) normalizes the distance value on the string length and can be formalized as:

$$fHD(r_i, r_j) = \frac{1}{N} \sum_{n=0}^{N-1} \left(r_{i,n} \oplus r_{j,n}\right) \tag{10.5}$$

It returns a value in the range $[0,\ 1]$, yielding the value 0 if $r_i = \overline{r_j}$ and the value 1 if $r_i = r_j$.

Globally, for a population of R devices, we can estimate the uniqueness, also called *inter-chip HD*, averaging fHD calculated for all $\binom{R}{2}$ responses pairs:

$$Uniqueness = \frac{2}{R(R-1)} \sum_{i=1}^{R-1} \sum_{j=i+1}^{R} fHD(r_i, r_j) \times 100\,\% \tag{10.6}$$

If PUFs provide uniformly distributed and independent response bits, the global uniqueness turns out to be close to 50 % on average. Values higher or lower than 50 % are symptoms of lower chip distinguishability.

10.3.2 Reliability

Ideally, a PUF should always be able to exactly reproduce the same response when the same challenge is applied. However, as anticipated before, since PUFs are based on variations of electrical characteristics, a number of response bits might change under either stable or variable environmental conditions, such as temperature and power supply. To this aim, the stability metric can be used to estimate the percentage number of bits in a response which change value among responses obtained from a repeatedly applied challenge. For a device d, let M be the number of measurements of N-bit responses, $r'_{d,j}$ $(j = 1, 2, \ldots, m)$, and r_d be the baseline reference response of the d-th device. The stability, also called as *intra-chip Hamming Distance* or Steadiness in [16], can be estimated as:

$$\text{Stability}\ (i) = \frac{1}{M} \sum_{j=0}^{M-1} fHD\left(r_i, r'_{i,j}\right) \times 100\,\%. \tag{10.7}$$

For a population of R devices, it can be averaged as

$$\text{Stability} = \frac{1}{R} \sum_{i=0}^{R-1} \mathit{Stability}\ (i)\,. \tag{10.8}$$

Alternatively, we can express the reliability value, which is the percentage number of bits which keep the value stable over time

$$\text{Reliability}\ (d) = 100 - \mathit{Stability}\ (d) = \left(1 - \frac{1}{M} \sum_{j=1}^{M} fHD\left(r_d, r'_{d,j}\right)\right) \times 100\,\%.$$

A PUF with stable responses achieves a high value of reliability, thus its value should be as close as possible to 100 %.

10.3.3 Uniformity

A PUF is expected to generate responses containing ideally the same number of logic-0s and logic-1s. Therefore, the uniformity metric (also called randomness by Yu et al. in [47]) can be exploited to estimate the distribution of logic-0 and logic-1 in PUF responses. Let N be the number of response bits, the percentage measure for uniformity of response $r_i = (r_{i,0},\ r_{i,1},\ \ldots\ r_{i,N-1})$ can be defined as

$$\text{Uniformity}\ (i) = \frac{1}{N} \sum_{n=0}^{N-1} r_{i,n} \times 100\,\%. \tag{10.9}$$

A value of 100 % means that all r_i response bits are logic-1. For true random bits, uniformity should be as close as possible to its ideal value of 50 %. Let R be the number of responses, resulting from the product between the amount of different

PUF instances and input challenges (if any). The average uniformity for a population or R devices can be calculated as

$$\text{Uniformity} = \frac{1}{R}\sum_{i=1}^{R} \mathit{Uniformity}\,(i) \tag{10.10}$$

10.3.4 Bit Aliasing

The uniformity metric is not enough to qualify the randomness of PUFs responses. Indeed, even with a best value of bit uniformity, some homologous bits could turn out to be biased among the responses of the PUF population. This could happen whenever the manufacturing process introduces static variations which compromise all the homologous bits in the responses, causing a fixed preferred value. To this aim, we can compute the bit aliasing [30] (also called bias in [47]) as

$$\text{Bit-Aliasing}\,(n) = \frac{1}{N}\sum_{i=0}^{R-1} r_{i,n} \times 100\,\%, \ \forall n \tag{10.11}$$

If some homologous bits are biased, the bit-aliasing results in a value far from 50 %.

10.4 Categories of PUFs

This section tries to cover all architectures of intrinsic silicon PUFs which have been proposed in the literature, giving a complete overview trough a categorization. The adjective intrinsic specifies that the PUF and related measurement circuits are manufactured on the same chip and using technological primitives which are inherently available in the target implementing technology. Mainly we can list PUFs by their operational mechanism, being some architectures based on a delay measurement (Sect. 10.4.1) and the others based on the start-up value of memory cells (Sect. 10.4.2).

10.4.1 Delay-Based PUF

PUFs which exploit delay measurement to extract responses are categorized as delay based. We can figure out a delay-based PUF as a digital contest between two paths which have to be resolved by a circuit, so-called arbiter, which decided which path has won the contest. Paths need to be designed symmetrically such that they ideally are characterized by the same delay. This way, the winner cannot be early established, but only once the chip is manufactured. Indeed, variations introduced during the production process modify the physical parameters of the chip in a random fashion and decide the exact delay that characterizes each path and, hence, the outcome of each PUF circuit.

10.4.1.1 Arbiter PUF

The arbiter PUF was introduced by Lim et al. in [25] and its mechanism is based on the previously introduced delay contest. To have more paths available for a comparison, authors proposed a scheme composed of a cascade of switch blocks, reported in Fig. 10.1. Each switch block can propagate an input signal through two configurable path, i.e., connected straight or switched. A configuration bit establishes in which configuration of the block has to work. An arbiter, namely a flip-flop, produces an output which depends on the difference between the two propagation delays: logic-1 if the signal which drives the data input (D) is faster than the clock signal, otherwise logic-0. The challenges set cardinality exponentially increases with the number of switch blocks: with N switch blocks, the circuit is able to compare 2^N different paths pairs.

The circuit could suffer from metastability condition if the offset between delays is close to be 0: such time violation will cause random behavior because the output is not deterministic.

10.4.1.2 Ring Oscillator PUF

Differently from the Arbiter PUF, the ring oscillator PUF (RO PUF) evaluates delay differences through frequency measurements exploiting asynchronous oscillating loops [41]. Each loop contains an odd number of inverting stages, such that, once the circuit is powered on, it starts to indefinitely oscillate. A generic controlled ring

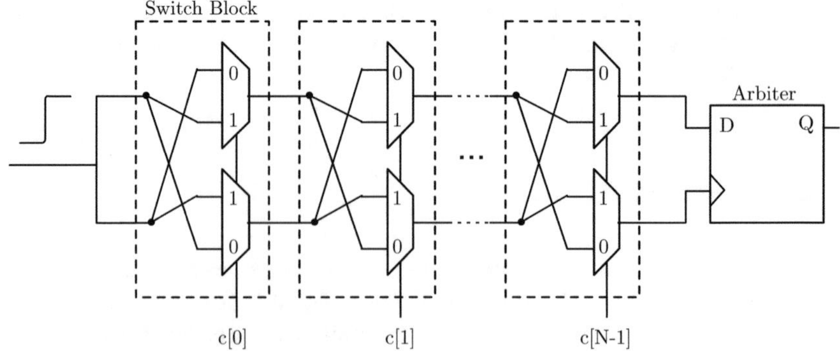

Fig. 10.1 Schematic of the arbiter PUF, originally proposed in [25]

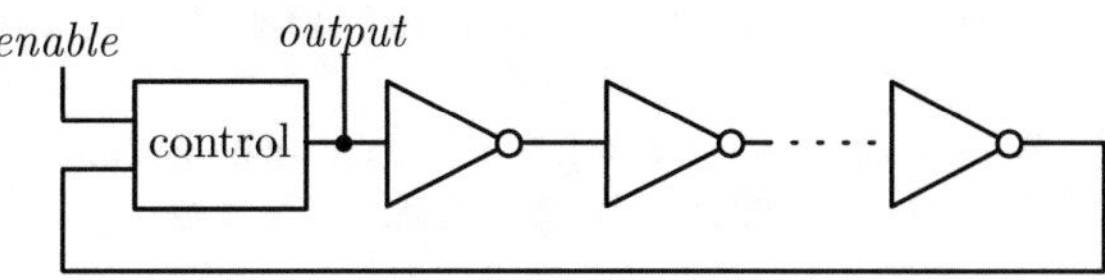

Fig. 10.2 Controlled ring oscillator schematic

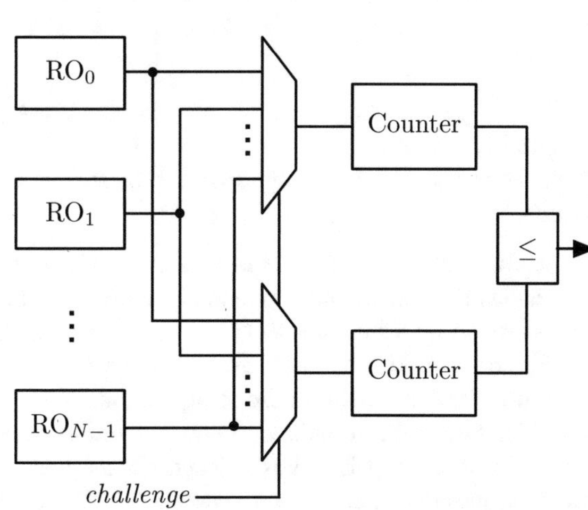

Fig. 10.3 Controlled ring oscillator schematic

oscillator is reported in Fig. 10.2: the output of the last stage of the ring is fed back into a control gate, together with the enable signal. If the number of inverting stages in the loop is even, the control gate is an *and*, otherwise is a *nand*. The loop can be also composed by only one inverting stages and other delay elements, such as non-inverting buffers [5].

Contrary to the arbiter PUF, which requires a perfect symmetry of two delay lines, the RO PUF requires ring oscillator circuits equally laid-out. For this reason, the RO PUF can be easily realized on every silicon technology, including FPGA.

The RO PUF is composed of N identically laid-out ring oscillators, two multiplexers that select a pair among them, and two counters that measure frequencies of the pair. Due to the manufacturing imperfections, the frequency of oscillations of each loop is random and device dependent, hence it slightly differs from the ideal frequency value. The frequency measurement also involves another counter, driven by the system clock, which establishes the time window in which the oscillations edges have to be counted.

A challenge decides which pair of ring oscillator has to be measured. Once the frequencies are extracted, a circuit compares them giving logic-1 or logic-0. The full pairing strategy allows to get $\frac{N(N-1)}{2}$ distinct ring oscillator pairs to compare, however the entropy that the circuit is able to generate is less than $\frac{N(N-1)}{2}$, since some bits obtained by frequencies comparisons are correlated. To maximize the entropy, the pairing strategies has to pick only $\frac{N}{2}$ (Fig. 10.3).

In [48], Yu et al. have introduced the ring oscillator sum PUF, which exploits k ring oscillators pairs, each one characterized by a specific frequency difference.

Once delays are measured, they are summed and the challenge bits determine the sign of each delay.

A more complex pairing strategy has been introduced by Yin and Qu in [45], which exploits a sequential pairing algorithm that generates reliable bits. In particular, the proposal gives $\frac{N}{2}$ responses bits from N ring oscillators Another pairing strategy, the chain-like neighbor coding, has been proposed in [46]. It consists of two design principles:

1. ROs have to be placed as close to each other as possible to minimize systematic variations effect;
2. pairing has to consider only adjacent ring oscillator, hence it generates only $N - 1$ bits.

Testing this strategy on the FPGAs leads to an improvement in uniqueness.

Besides the pairing strategy and challenge size, the ring oscillator frequency is characterized by a high susceptibility to environmental changes, such as die temperature and working voltage, that could cause instability in responses, especially for pairs which frequencies are really similar (Fig. 10.4). So far this problem has arosed and it has been addressed with a 1-out-of-k masking scheme [41], that has the aim of picking pairs whose distance is the maximum one among k pairs. The overhead introduced can be estimated as $\frac{2}{k} \cdot 100\,\%$, since the architecture requires only 2 of k pairs to extract a response bit. Such a redundancy can be implemented on a Xilinx FPGA maintaining the same area overhead by using the configurable ring oscillator [31]. In the configurable structure the loop contains six inverters selected by thre multiplexers, which are able to define eight different ring oscillators. The area overhead of this configuration is the same of a normal ring oscillator, since the configurable architecture occupies the whole CLB.

Furthermore, it has been proven that the logic which surrounds ring oscillators is able to alter frequencies values. Merli et al. in [35] empirically demonstrated a relation between the spatial frequency distribution and the position and shape of surround logic, synthesized to read out the frequency of ring oscillator, using Xilinx Spartan-3E and 2712 ring oscillators instantiated by means of hard macro. In particular, due to local effect, such as different current flows and temperature, can be changed by other active logic which is working on chip together with ring oscillator. Moreover, for a single ring oscillator, the other ones represent surrounding logic, hence the authors proposed to enable/disable them in order to have only one working at time.

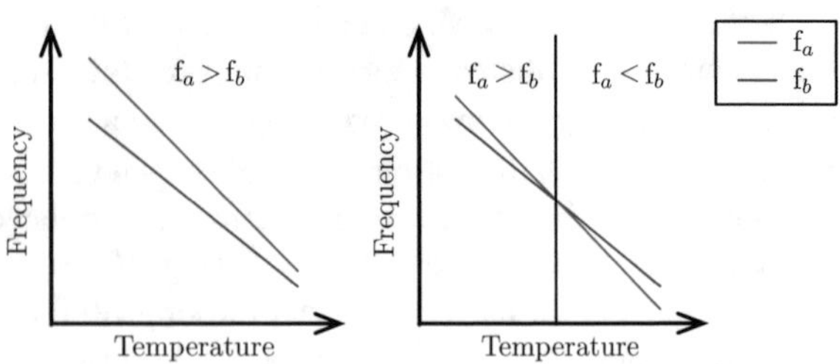

Fig. 10.4 Frequencies variation against temperature when they are enough far apart and when they are close each other

Other measurement campaigns have been reported in [13, 30], respectively, on Altera Cyclone IV and Xilinx Spartan-3E.

10.4.1.3 Anderson PUF

The Anderson PUF, introduced the first time in [2], is a PUF architecture devised to be synthesized on Xilinx Virtex-5 devices, as primitives which exploits are specific for this technology. The Anderson PUF is a composition of basic elements, defined as Anderson cells, and each one outputs only 1-bit PUF response. The cell contains two shift registers, two 2-to-1 multiplexers and one D flip-flop. The 0 data input of multiplexers is stuck to logic-0, the select input is driven by the outputs of shift registers, which are synchronous with the same clock signal and generate a counterphase logic-1 and logic-0 sequences. Multiplexer B has its 1 data input tied to logic-1 and its output $N1$ is connected to the 1 data input of the multiplexer A. Figure 10.5 details all the connections within the Anderson PUF cell. Due to manufacturing variation, outputs of shift registers are not perfectly in counterphase, but there are some overlaps during the switching that generate positive glitches. As be changed to logic-1 from logic-0 by a positive glitch that appears on the preset port. Furthermore, the flip-flop is configured as one-catcher.

Shift operations of the Anderson PUF are implemented using two *SLICEM* LUTs, configured as 16-bit shift registers. The two multiplexers are interconnected using the carry chain, and the flip-flop can be located in the same slice of the other elements. Therefore, the Anderson PUF cell could be theoretically implemented using only one slice, but glitch modulation has to be considered. To this aim, Anderson increased the glitch pulse width by varying the distance between the two multiplexers. The best configuration in terms of responses quality was composed of five intermediate carry chain multiplexers.

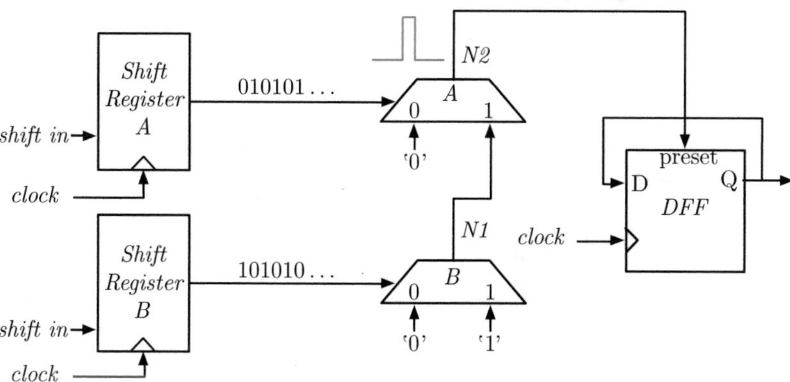

Fig. 10.5 Logic schematics of the Anderson PUF cell

Experimental results showed that the PUF responses have a good behavior when the working temperature changes, i.e., the PUF stability has good value to be used as a secure primitive on the FPGA.

The original proposal on the Virtex-5 has been implemented also on other Xilinx devices family. In [18] Huang and Li proposed an implementation on the Xilinx Virtex-6, integrating also a mechanism to provide one challenge bit in the cell. Moreover, the authors configured the shift registers with a different initialization sequences (0x8888 and 0x4444 instead of 0xAAAA and 0x5555). In [49] Zhang et al. detailed an implementation of the Anderson PUF on Xilinx Zynq-7000.

10.4.1.4 ALU PUF

The ALU PUF, introduced in [21] by Kong et al., is a delay-based PUF which exploits two symmetric arithmetic logic units (ALUs). Due to delay mismatches introduced by manufacturing variability, ALUs execute operations with different time. This is also true for the carry chain of ripple carry adders, a fundamental circuit for ALUs. A high level architectural schematic of a 4-bit ALU PUF is pictured in Fig. 10.6. Querying ALUs with same operands (challenge $x_0, \ldots, x_7$) and evaluating the differences in propagation delay, make possible to generate response bits $(y_0, \ldots, y_3)$ with a minimal hardware overhead, namely synchronization logic, to ensure that inputs

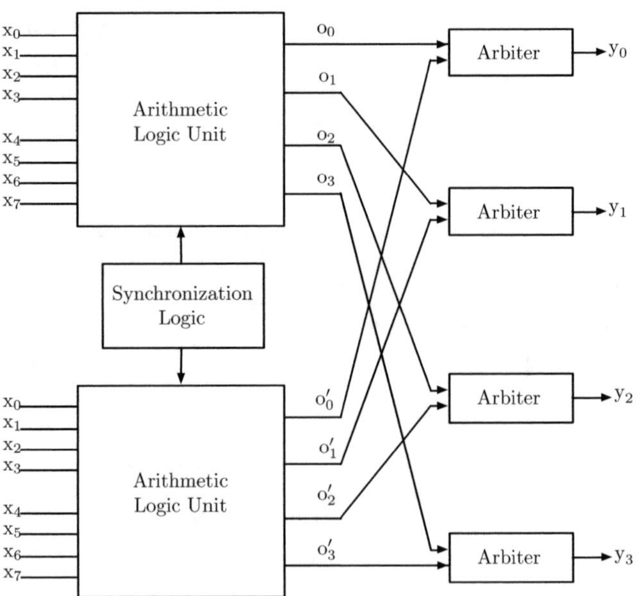

Fig. 10.6 Schematic of the ALU PUF architecture

reach both the ALUs at the same time, and arbiters, which evaluate mismatches in signal delays to extract one response bits.

Depending on the bit-operand parallelism, the ALU PUF can be realized with an arbitrary number of response bits.

10.4.2 Memory-Based PUF

A memory-based PUF uses the random initial state of a memory cell on a device start-up to extract a signature which is device dependent. Contrary to the previous introduced delay-based PUFs, memory PUFs do not require specific design to generate a signature, but they work with standard memory cell.

10.4.2.1 SRAM PUF

The static random access memory (SRAM) PUF, exploits the start-up value of SRAM cells, which are memory blocks realized with two cross-coupled inverters, to generate a unique and unclonable fingerprint [23]. In CMOS technology, as shown in Fig. 10.7a, each cell requires six transistors and the structure is symmetrically realized with two halves. During the start-up, each cell reaches and keeps the initial state even if the transistors are nor directly driven by external signals. Once powered, the two halves are characterized by an unstable voltage point and, due to small variations introduced by manufacturing process, they force each other to reach one of two possible stable points, low or high state, by amplifying the differences of voltages, as depicted in Fig. 10.7. As discussed in [10], there are three types of SRAM cells: (i) non-skewed cells, that are characterized by a random behavior, since the effect of manufacturing variations on the two halves turn out to be mutually neutralized;

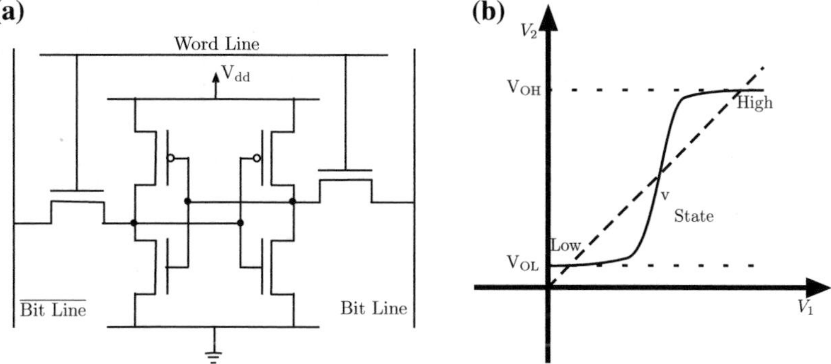

Fig. 10.7 Details on SRAM cell. **a** Transistor-level schematic of a SRAM cell in CMOS technology. **b** Input–output voltage graph of cross coupled inverting stages

(ii) partially skewed cells, that have a mismatch between the two inverters such that they have a preferred state, but external conditions might provoke a state flip; (iii) fully skewed cells, like the partially skewed, are characterized by a preferred state which does not change, even under different external conditions. Maes et al. in [28] illustrated an approach to implement a low-overhead algorithm to obtain high SRAM PUF response stability.

Guajardo et al. adopted the SRAM of an FPGA with embedded block RAM memories to extract unique and unclonable responses. The main problem with the FPGA technology is the programming phase, since when the configuration phase initialized the memory block to fixed values. In [44] the authors reported the power gating feature of Xilinx FPGAs to extract the start-up value from block RAM exploiting the dynamic partial reconfiguration. Other non-FPGA SRAM tests have been reported in [4, 39, 40].

10.4.2.2 Butterfly and Flip-Flop PUF

In order to have available memory-based PUF on FPGA and address the drawback of the SRAM PUF, Kumar et al. introduced the butterfly PUF, which is a PUF based on cross-coupled memory elements [22]. In particular, they adopted two cross-coupled transparent latches configured as reported in Fig. 10.8. Each latch has a preset and a clear input, that work asynchronously. Preset of one latch and clear of the other one are driven by the same signal, called excite, while the D input is driven by the Q of the other latch. When the excite signal is asserted, the circuit is in an unstable state due to the latches that have opposite states. Ideally, the circuit should indefinitely retain this state, but once the excite is set to low, the circuit tends to one of the two

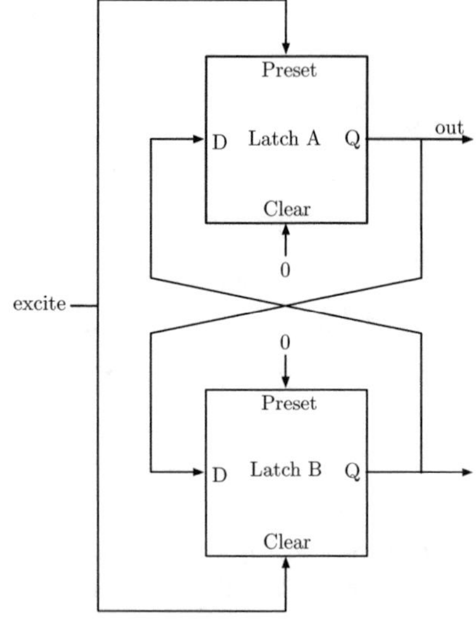

Fig. 10.8 Schematic of the butterfly PUF

stable states due to physical mismatches. The reached states can be used as a bit of PUF response.

Equivalently to the SRAM PUFs, Maes et al. proposed to read the start-up values of flip-flops on FPGAs in [27]. The flip-flop initialization value is established by the configuration contained in the bitstream: this value is not immediately loaded into the flip-flops, but only after the complete loading of the new configuration onto the FPGA. Indeed, once the configuration is accomplished, the global restore line is asserted, causing the reset of flip-flops to the value specified in the bitstream. Basically, the idea of the Flip-Flop PUF is based on the disabling of the reset logic of regular flip-flops when the device is programmed.

10.4.2.3 STT-MRAM PUF

SRAM PUFs are one of the most investigated solutions, since, as anticipated before, they do not require additional hardware overhead to extract unique and unclonable fingerprints from integrated circuits. Moreover, the quality and stability of responses makes them really attractive, compared with other PUF architectures. Recently, the focus is moving toward emerging memory technologies, such as magnetic resistive memories. Vatajelu et al. proposed an innovative PUF architecture in [43], which exploits the manufacturing variability of Spin-Transfer Torque Magnetic RAM. Indeed, the electrical resistance of magnetic tunnel junction in antiparallel magnetization highly suffer from imperfection induced by the manufacturing process.

The Fig. 10.9 details a STT-MRAM cell circuit and the configuration in which it stores the logic-0 and the logic-1 value. To extract the stored value, hence to evaluate the MTJ resistance, a reference current involves in the process and it is generated by reference cells. Active and reference cells are identically manufactured.

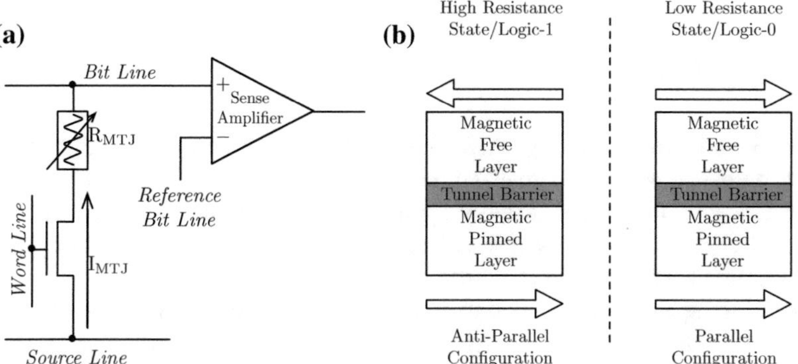

Fig. 10.9 The STT-MRAM cell. **a** Electrical equivalent circuit of a one transistor and one MTJ structure.**b** MTJ configurations of logic-0 and logic-1 states

The PUF takes advantage of the variability of MTJ resistances in antiparallel configuration. To extract a unique fingerprint, all cells are configured in antiparallel state (writing logic-1 values) and the active ones are read through the sense amplifier, comparing their current with the current of reference cells. The state of active cells will be interpreted as logic-0 if $I_{MTJ} > I_{Ref}$, otherwise as logic-1 if $I_{MTJ} < I_{Ref}$. Even if all the cells are configured in logic-1 state, being the MTJ resistances normally distributed, statistically half of them will be interpreted as logic-0.

10.5 Post-Processing Techniques

Unclonability and unpredictability properties turn PUFs really sound for key material provider and key storage circuit. Indeed, keys are the critical point for any crypto-system and they must be protected against unauthorized accesses. Contrary to memory registers, PUFs can provide keys on demand and only when the circuit is powered up, so keys reside nowhere. Moreover, they cannot be cloned and predicted, and are safeguarded against physical attacks thanks to the anti-tamper properties of PUFs [24].

Despite such advantages, PUF responses cannot be directly used as cryptographic keys due to the noise generated by changes in environmental conditions. Therefore, a reliability algorithm is added as post-processing technique to the PUF architecture to deal with responses noise, generated when working conditions, such as temperature, voltage, and device aging, change between the provisioning reference, i.e., during the enrollment phase, and the conditions during the regeneration phase, which occurs in field.

Post-processing technique also accounts for other quality parameters, such as uniformity of information entropy.

10.5.1 Majority Voter

Majority voting is an effective technique when PUF responses are characterized by low or transient noise. It is realized through the collection of a significant number of responses: if they are repeatedly extracted from the same PUF, the technique is defined as temporal majority voting, vice versa if they are taken from multiple PUFs at once, the method is defined as spatial majority voting [27].

As for the spatial majority voting, its discrete nature impedes to reach good values of bits uniformity in responses. As for the temporal majority voting, the technique needs for a number of repeated measures which exponentially increase with the desired noise reduction, due to the Chernoff bound.

To be used as keys, PUF responses have to guarantee perfect distribution of bits and noise-free responses.

10.5.2 Fuzzy Extractor

Contrary to majority voting algorithms, fuzzy extractor schemes involve an error-correction code algorithm to set PUF responses free from the noise and require to collect only one sample per single response [12, 26]. Typically, a fuzzy extractor scheme requires two main phases. The first one, namely *generation phase*, the PUF is enrolled and the obtained response is securely stored. Moreover, an additional bit string, called *helper data*, is generated. The second one, called *reproduction phase*, exploits the previously defined *helper data* in order to recover noisy version of the enrolled response, The *helper data* are not critic for the scheme and can be publicly exchanged, as they do not weaken secrecy of the PUF response. If the noise which affects the response is small enough, the *reproduction phase* guarantees that the reproduced response perfectly matches the PUF response extracted in the first phase. In particular, depending on the error-correction design parameters, fuzzy extraction is able to recover a response if the amount of error bits is under a certain threshold. Figure 10.10 illustrates every step for generation and reproduction phases.

Besides the error correction, a fuzzy extractor scheme comprises also an additional step, called privacy amplification. First of all, it is used to extract random bits such that the PUF response turns out enhanced in uniformity (see Sect. 10.3.3), gaining information entropy, which is necessary whenever responses have to be used as secure keys [1, 15].

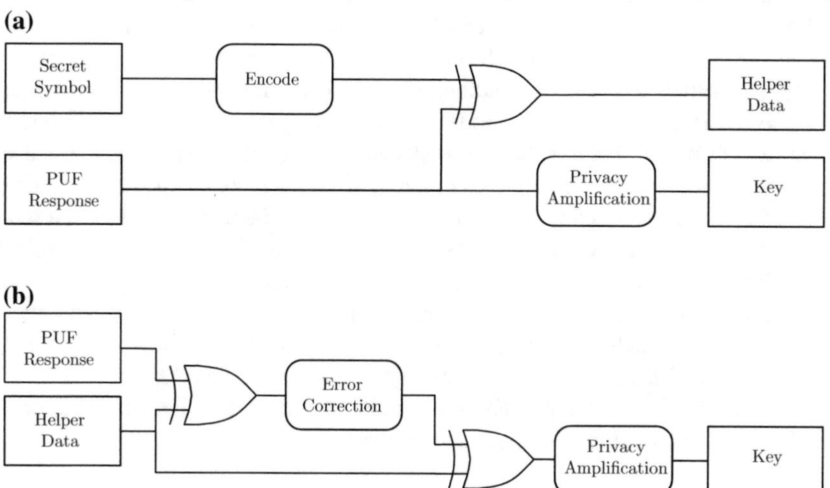

Fig. 10.10 The fuzzy extraction algorithm scheme. **a** Generation phase, to accomplish during the PUF enrollment in order to extract the key and the *helper* data. **b** Reproduction phase, which recover a noise version of the response and outputs the same key extracted in generation

10.6 Attacks Against PUF

The adoption of a PUF as key storage and generator avoids some of the shortcomings of memory-based approaches. It is generally harder to perform an attack aimed to read out, predict, or derive responses from a PUF than to gain access to a nonvolatile memory. However, PUF implementations may be subject of attacks. They could exploit either the working model of a PUF circuit or post-processing algorithms.

10.6.1 Model Based Attack

The model building attack can be accomplished by collecting a large number of CRPs that lead to extraction of a mathematical function of the PUF under attack, such that an attacker is able to predict responses generated by arbitrary-defined challenges [3, 14, 24, 38]. Only PUFs with a large set of CRPs, namely, strong PUF [15], are prone to such attack.

Modeling attacks may involve machine learning algorithms, which are suitable to extract enough knowledge from a subset of CRPs and to generalize, i.e., predict, the mapping mechanism behind a PUF. The extracted model is a digital clone of the PUF and violates the mathematical unclonability property (see Eq. 10.2a). Anyway, to be successful the machine learning tool has to determine a polynomial-sized timing model which is as much as possible accurate in predict responses.

Without loss of generality, we can consider a delay PUF. Assuming that each element delay that belongs to a path add up itself to the total path delay, an adversary can apply a sequence of inputs (challenges) to the PUF and obtain a system of equations, which expresses the linear model behind the mapping mechanism. Whenever the system turns out linear, solving this system is relatively simple, since it is a set of linear equations in the continuous domain, and consequently the support vector machines are a good machine learning tool candidate to be used. Otherwise, a more complex set of nonlinear equations has to be considered. Their complexity depends on the fact that the delay of each element delay is a nonlinear function of the previous element delay, whose delay strongly depends on the order of the transitions converging on it. Machine learning techniques can still be used, however the training set is nonlinearly separable, hence other algorithms have to be adopted.

As a matter of fact, Arbiter PUF implementations show an additive linear behavior, which makes them vulnerable to modeling attacks [17, 32, 38].

10.6.2 Side-Channel Attack

Side-channel attack exploits the non-primary outputs of an algorithm to gather information not available as primary output [19, 20]. Such information are leaked during the execution of the algorithm by measuring power consumption, execution time, emitted electromagnetic field, temperature, and so on. Attacks based on the side channel technique can be performed to extract PUF responses when they are processed to remove noise and recover quality parameters. Authors of [33, 34] successfully performed and documented two types of side-channel attacks to a RO PUF by exploiting the fuzzy extractor which was implemented to post-process the responses. Similarly, authors of [11] accomplished an attack against an error-correction code for a weak PUF (a PUF that is not strong).

10.7 Conclusion

In this chapter, we tried to give a rigorous overview on the PUF research field, with particular attention to existing architectures and adopted quality metrics. Indeed, as PUF is a concept which has been formed during time, we gave a collection of functional aspects with a formal notation. On the pragmatic level, we included in the chapter an exhaustive list of proposed PUF architectures in the literature, illustrating their strong points and main characteristics. Moreover, we briefly reported some attacks against PUFs, covering all categories of attacks.

References

1. Amelino D, Barbareschi M, Battista E, Mazzeo A. How to manage keys and reconfiguration in WSNs exploiting sram based PUFs. In: Intelligent interactive multimedia systems and services 2016. Springer International Publishing; 2016. p. 109–19.
2. Anderson JH. A PUF design for secure FPGA-based embedded systems. In: Proceedings of the 2010 Asia and South Pacific design automation conference. IEEE Press; 2010. p. 1–6.
3. Barbareschi M, Bagnasco P, Mazzeo A. Authenticating IOT devices with physically unclonable functions models. In: 2015 10th international conference on P2P, parallel, grid, cloud and internet computing (3PGCIC). IEEE; 2015. p. 563–7.
4. Barbareschi M, Battista E, Mazzeo A, Mazzocca N. Testing 90 nm microcontroller SRAM PUF quality. In: 2015 10th international conference on Design and Technology of Integrated Systems in Nanoscale Era (DTIS). IEEE; 2015. p. 1–6.
5. Cheri Z, Danger J-L, Guilley S, Bossuet L. An easy-to-design PUF based on a single oscillator: the loop PUF. In: 2012 15th euromicro conference on Digital System Design (DSD). IEEE; 2012. p. 156–62.
6. Cilardo A. Efficient bit-parallel GF(2^m) multiplier for a large class of irreducible pentanomials. IEEE Trans Comput. 2009;58(7):1001–8.
7. Cilardo A. New techniques and tools for application-dependent testing of FPGA-based components. IEEE Trans Ind Inform. 2015;11(1):94–103.
8. Cilardo A, Barbareschi M, Mazzeo A. Secure distribution infrastructure for hardware digital contents. IET Comput Digit Tech. 2014;8(6):300–10.
9. Cilardo A, Mazzocca N. Exploiting vulnerabilities in cryptographic hash functions based on reconfigurable hardware. IEEE Trans Inform Forensics Secur. 2013;8(5):810–20.

10. Cortez M, Dargar A, Hamdioui S, Schrijen G-J. Modeling sram start-up behavior for physical unclonable functions. In: 2012 IEEE international symposium on defect and fault tolerance in VLSI and nanotechnology systems (DFT). IEEE; 2012. p. 1–6.
11. Dai J, Wang L. A study of side-channel effects in reliability-enhancing techniques. In: DFT'09. 24th IEEE international symposium on defect and fault tolerance in VLSI systems, 2009. IEEE; 2009. p. 236–44.
12. Dodis Y, Reyzin L, Smith A. Fuzzy extractors: how to generate strong keys from biometrics and other noisy data. In: Advances in cryptology-eurocrypt 2004. Springer; 2004. p. 523–40.
13. Feiten L, Spilla A, Sauer M, Schubert T, Becker B. Analysis of ring oscillator PUFs on 60 nm FPGAs. In: European cooperation in science and technology.
14. Gassend B, Clarke D, Van Dijk M, Devadas S. Silicon physical random functions. In: Proceedings of the 9th ACM conference on computer and communications security. ACM; 2002. p. 148–60.
15. Guajardo J, Kumar SS, Schrijen G-J, Tuyls P. Physical unclonable functions and public-key crypto for FPGA ip protection. In: International conference on field programmable logic and applications, 2007. FPL 2007. IEEE; 2007. p. 189–95.
16. Hori Y, Yoshida T, Katashita T, Satoh A. Quantitative and statistical performance evaluation of arbiter physical unclonable functions on FPGAs. In: 2010 international conference on reconfigurable computing and FPGAs (ReConFig). IEEE; 2010. p. 298–303.
17. Hospodar G, Maes R, Verbauwhede I. Machine learning attacks on 65 nm arbiter PUFs: accurate modeling poses strict bounds on usability. In: 2012 IEEE international Workshop on Information Forensics and Security (WIFS). IEEE; 2012. p. 37–42.
18. Huang M, Li S. A delay-based PUF design using multiplexers on FPGA. In: 2013 IEEE 21st annual international symposium on Field-Programmable Custom Computing Machines (FCCM). IEEE; 2013. p. 226.
19. Kocher P, Jaffe J, Jun B. Differential power analysis. In: Advances in CryptologyCRYPTO99. Springer; 1999. p. 388–97.
20. Kocher PC. Timing attacks on implementations of Diffie-Hellman, RSA, DSS, and other systems. In: Advances in CryptologyCRYPTO96. Springer; 1996. p. 104–13.
21. Kong J, Koushanfar F, Pendyala PK, Sadeghi A-R, Wachsmann C. PUFatt: embedded platform attestation based on novel processor-based PUFs. In: 2014 51st ACM/EDAC/IEEE Design Automation Conference (DAC). IEEE; 2014. p. 1–6.
22. Kumar SS, Guajardo J, Maes R, Schrijen G-J, Tuyls P. The butterfly PUF protecting ip on every FPGA. In: IEEE international workshop on hardware-oriented security and trust, 2008. HOST 2008. IEEE; 2008. p. 67–70.
23. Layman PA, Chaudhry S, Norman JG, Thomson JR. Electronic fingerprinting of semiconductor integrated circuits, May 18 2004. US Patent 6,738,294.
24. Lee JW, Lim D, Gassend B, Suh GE, Van Dijk M, Devadas S. A technique to build a secret key in integrated circuits for identification and authentication applications. In: 2004 symposium on VLSI circuits, 2004. Digest of technical papers. IEEE; 2004. p. 176–9.
25. Lim D, Lee JW, Gassend B, Suh GE, Van Dijk M, Devadas S. Extracting secret keys from integrated circuits. IEEE Trans Very Large Scale Integr VLSI Syst. 2005;13(10):1200–5.
26. Linnartz J-P, Tuyls P. New shielding functions to enhance privacy and prevent misuse of biometric templates. In: Audio-and video-based biometric person authentication. Springer; 2003. p. 393–402.
27. Maes R, Tuyls P, Verbauwhede I. Intrinsic PUFs from flip-flops on reconfigurable devices. In: 3rd Benelux workshop on information and system security (WISSec 2008), vol. 17; 2008.
28. Maes R, Tuyls P, Verbauwhede I. Low-overhead implementation of a soft decision helper data algorithm for SRAM PUFs. In: Cryptographic hardware and embedded systems-CHES 2009. Springer; 2009. p. 332–47.
29. Maes R, Verbauwhede I. Physically unclonable functions: a study on the state of the art and future research directions. In: Towards hardware-intrinsic security. Springer; 2010. p. 3–37.
30. Maiti A, Casarona J, McHale L, Schaumont P. A large scale characterization of RO-PUF. In: 2010 IEEE international symposium on Hardware-Oriented Security and Trust (HOST). IEEE; 2010. p. 94–9.

31. Maiti A, Schaumont P. Improved ring oscillator PUF: an FPGA-friendly secure primitive. J Cryptol. 2011;24(2):375–97.
32. Majzoobi M, Koushanfar F, Potkonjak M. Testing techniques for hardware security. In: IEEE international test conference, 2008. ITC 2008. IEEE; 2008. p. 1–10.
33. Merli D, Schuster D, Stumpf F, Sigl G. Semi-invasive em attack on FPGA ro PUFs and countermeasures. In: Proceedings of the workshop on embedded systems security. ACM; 2011. p. 2.
34. Merli D, Schuster D, Stumpf F, Sigl G. Side-channel analysis of PUFs and fuzzy extractors. In: Trust and trustworthy computing. Springer; 2011. p. 33–47.
35. Merli D, Stumpf F, Eckert C. Improving the quality of ring oscillator PUFs on FPGAs. In: Proceedings of the 5th workshop on embedded systems security. ACM; 2010. p. 9.
36. Naccache D, Fremanteau P. Unforgeable identification device, identification device reader and method of identification, July 18 1995. US Patent 5,434,917.
37. Rampon J, Perillat R, Torres L, Benoit P, Di Natale G, Barbareschi M. Digital right management for IP protection. In: 2015 IEEE computer society annual symposium on VLSI (ISVLSI). IEEE; 2015. p. 200–3.
38. Rührmair U, Sehnke F, Sölter J, Dror G, Devadas S, Schmidhuber J. Modeling attacks on physical unclonable functions. In: Proceedings of the 17th ACM conference on computer and communications security. ACM; 2010. p. 237–49.
39. Schrijen G-J, van der Leest V. Comparative analysis of SRAM memories used as PUF primitives. In: Design, Automation Test in Europe conference exhibition (DATE); 2012. p. 1319–24.
40. Selimis G, Konijnenburg M, Ashouei M, Huisken J, De Groot H, Van der Leest V, Schrijen G-J, Van Hulst M, Tuyl P. Evaluation of 90 nm 6T-SRAM as physical unclonable function for secure key generation in wireless sensor nodes. In: 2011 IEEE International Symposium on Circuits and Systems (ISCAS). IEEE; 2011. p. 567–70.
41. Suh GE, Devadas S. Physical unclonable functions for device authentication and secret key generation. In: Proceedings of the 44th annual design automation conference. ACM; 2007. p. 9–14.
42. Tolk KM. Reflective particle technology for identification of critical components. Technical report, Sandia National Labs., Albuquerque, NM (United States); 1992.
43. Vatajelu EI, Natale GD, Barbareschi M, Torres L, Indaco M, Prinetto P. STT-MRAM-based PUF architecture exploiting magnetic tunnel junction fabrication-induced variability. J. Emerg. Technol. Comput. Syst. 2016;13(1):5:1–5:21. http://doi.acm.org/10.1145/2790302.
44. Wild A, Guneysu T. Enabling SRAM-PUFs on xilinx FPGAs. In: 2014 24th international conference on Field Programmable Logic and Applications (FPL). IEEE; 2014. p. 1–4.
45. Yin CED, Qu G. LISA: maximizing RO PUF's secret extraction. In: 2010 IEEE international symposium on Hardware-Oriented Security and Trust (HOST). IEEE; 2010. p. 100–5.
46. Yin C-E, Qu G. Improving PUF security with regression-based distiller. In: Proceedings of the 50th annual design automation conference. ACM; 2013. p. 184.
47. Yu M-D, Sowell R, Singh A, M'Raihi D, Devadas S. Performance metrics and empirical results of a PUF cryptographic key generation ASIC. In: 2012 IEEE international symposium on Hardware-Oriented Security and Trust (HOST). IEEE; 2012. p. 108–15.
48. Yu MDM, Devadas S. Recombination of physical unclonable functions; 2010.
49. Zhang J, Wu Q, Lyu Y, Zhou Q, Cai Y, Lin Y, Qu G. Design and implementation of a delay-based PUF for FPGA IP protection. In: 2013 international conference on Computer-Aided Design and Computer Graphics (CAD/Graphics). IEEE; 2013. p. 107–14.

Chapter 11
Implementation of Delay-Based PUFs on Altera FPGAs

Linus Feiten, Matthias Sauer and Bernd Becker

11.1 Introduction

Altera is—besides Xilinx—the largest manufacturer of field-programmable gate arrays (FPGAs) and their devices are widely used. Over the years, there have been several variants of the Cyclone FPGA series. The first version based on 130 nm process technology was introduced in 2002, followed by the Cyclone II (90 nm, 2004), Cyclone III (65 nm, 2007), Cyclone IV (60 nm, 2009) and Cyclone V (28 nm, 2011). Despite advancements from version to version, general architectural concepts are sustained, setting Altera FPGAs apart from the Xilinx architecture. In the course of this chapter, the reader will be introduced to the Cyclone architecture and be enabled to put the concepts of delay-based PUFs [16] into practice. This is done using the Altera design software Quartus II and the hardware description language VHDL. The communication between the FPGA and a PC is done via the JTAG interface using the scripting language TCL and custom commands provided by Altera SignalTap II. Beforehand, however, a short summary of delay-based PUFs is given and their application in FPGAs as opposed to application-specific integrated circuits (ASICs) is put into context.

The purpose of a PUF is to have a unique signature (typically in form of a binary number) associated with each device, that is generated from the device's unique physical characteristics. This signature can be used to tell the device apart from other devices or even be part of a cryptographic protocol to allow only the device

L. Feiten (✉) · M. Sauer · B. Becker
University of Freiburg, Chair for Computer Architecture,
Georges-Koehler-Allee 51, 79110 Freiburg, Germany
e-mail: feiten@informatik.uni-freiburg.de

M. Sauer
e-mail: sauerm@informatik.uni-freiburg.de

B. Becker
e-mail: becker@informatik.uni-freiburg.de

N. Sklavos et al. (eds.), *Hardware Security and Trust*,
DOI 10.1007/978-3-319-44318-8_11

owner access to certain resources. Instead of a PUF generating such signature, it could simply be stored in the non-volatile memory of a device. However, an attacker might physically access this memory with comparatively little effort (e.g. [6, 17]). The advantage of a PUF is that the signature is only generated when needed and stored temporarily, making an attack much more difficult. Furthermore, as the PUF signature is generated from physical characteristics, an invasive attack on the device is likely to disturb these characteristics such that the signature is altered.

Whether the PUF signature can be read-out directly from the device or whether it remains concealed within it depends on the implementation. In the former case, an attacker might read-out the signature and possibly forge a device with the same signature; in the latter case—if the signature is only used inside of the device, e.g. as the seed for an asymmetric cryptographic key pair [18]—an attacker has to go through much greater efforts to possibly obtain the secret signature. Using the PUF output as a cryptographic seed, however, requires perfect reliability, because just a single bit-flip in the seed leads to a completely different key. As perfect reliability is hardly achieved in any PUF, error correction schemes [22] must be used. If the signature is read-out directly, non-perfect reliability can be compensated by considering a signature as "correct" when enough signature bits have their expected values.

To prevent an attacker from learning the whole signature of a device—intending to forge an identical device—so-called *strong* PUFs [15] have the potential of generating a signature of exponential length. To identify a device, only a subset of the whole signature is sampled, depending on a challenge. An attacker is hampered, because there are too many challenges to read out the complete signature, and he does not know which challenges are used by the legitimate owner to identify the device. *Weak* PUFs on the other hand, only generate a manageable amount of signature bits which could—if the signature is not concealed—be read out and stored by an attacker. Whether an attacker is able to forge the signature of a device depends on the technology and the PUF. Mostly, this will prove rather difficult as the physical device characteristics depend on uncontrollable variances in the production process. Only with extensive effort would it be possible to, e.g. alter the capacitances of single transistors that a PUF yields its responses from.

The implementation of PUFs on FPGAs—as opposed to ASICs—brings some peculiarities with it as their functionality is not hardwired. Instead, their reconfigurable hardware is configured to realise any feasible functionality being encoded in a so-called *bitstream* file. The bitstream is loaded to volatile memory elements of the FPGA any time it is powered up. We call such an FPGA configuration an FPGA *design*. During the creation of a design, the designer uses a hardware description language like VHDL or Verilog to describe the *netlist* defining all gates, memory elements and interconnects. The FPGA vendor's design software then maps those elements to the configurable FPGA hardware and compiles the distributable bitstream.

While the netlist and its mapping to the FPGA hardware is still "human-readable", the bitstream is definitely not. In fact, it is considered a security feature when the netlist cannot be reverse-engineered from the bitstream. FPGA vendors therefore make it a secret how their bitstream compilers really work and try to obscure how the bitstream relates to the actual FPGA configuration. However, there have been attempts to break this kind of "security by obscurity" [14], which is why some newest FPGA types provide bitstream encryption, where the bitstream is encrypted with a secret key that is also stored in a battery-powered memory of the FPGA.

There are several reasons why the bitstream should not be reverse-engineerable. The first being that the copyright owner of an FPGA design put a lot of research and development costs into its creation. While the netlist can be kept as a business secret, the bitstream cannot as it must be distributed in a file or non-volatile memory connected to the FPGA. Both can be easily accessed. When a business competitor manages to reverse-engineer a regularly purchased bitstream, he could reuse it for his own products evading the expensive research and development costs. Another scenario in which the bitstream should not be reverse-engineerable is the implementation of PUFs. Because if all placement and routing details of the PUF circuitry are known to an attacker, it is a lot easier for him to manipulate only the relevant hardware components to make the PUF produce another signature. Furthermore—in case of a concealed weak PUF—he could compile an FPGA design with just the same but non-concealed PUF.

Given that the bitstream cannot be reverse-engineered, a customer is able to programme an arbitrary amount of FPGAs with it. Such "overproduction" can be harmful to the business of an FPGA design vendor, who might want to sell bitstreams with licenses for limited usage. Bitstream encryption alone does not prevent this, as the customer must also possess the key to decrypt the bitstream. With a PUF, however, the vendor can link bitstreams to specific FPGAs by encoding the expected PUF signature into the logic of the design. Thus, the implemented hardware will only start functioning on an FPGA where the PUF produces the expected signature [23]. To link a bitstream to an FPGA, the vendor must have access to the customer's FPGA once in order to sample its unique signature.

*

There are many different kinds of PUF implementations on FPGAs; e.g. arbiter PUFs [9], butterfly PUFs [8], TERO-PUFs [2] or SRAM-PUFs [1, 7]. This chapter focuses on delay-based PUFs in general and how to implement them on Altera Cyclone FPGAs. The output of a delay-based PUF is determined by the delays of certain circuit lines. The delay determines how long a change from high voltage (logic 1) to low voltage (logic 0) or vice versa takes to travel through a conducting circuit line. In the production process of integrated circuits, unavoidable process variations lead to slightly different delays on each individual chip. For instance, the higher the resistance of a line, the greater its delay. Thus, these delays can be used as device-specific physical characteristics from which the PUF signature is generated. The delays are also highly dependent on operating temperature and voltage. Delay-based PUFs generally compensate such fluctuations by using the relative differences between circuit lines instead of absolute measurements.

A very popular kind of delay-based PUFs is the ring oscillator PUF (RO-PUF) that has received much research attention due to its relative simplicity and stability (e.g.

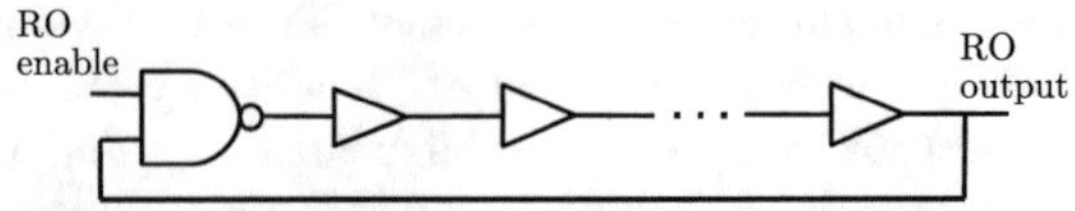

Fig. 11.1 A single ring oscillator (RO)

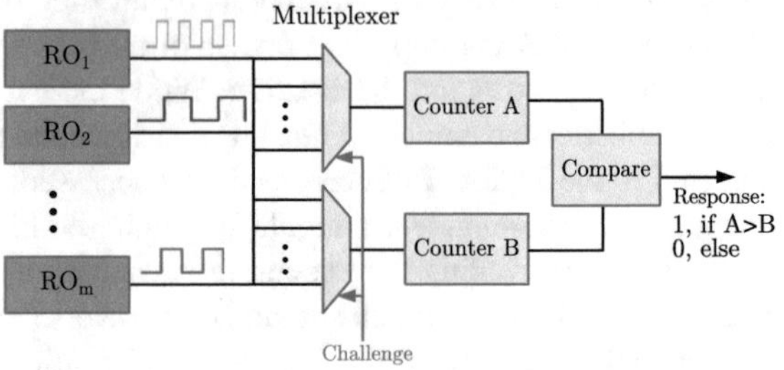

Fig. 11.2 An RO-PUF with *m* ROs, multiplexers, counters and comparators

[5, 10, 19–21]). We will therefore take it as the running example for this chapter, but the demonstrated methods can be used to implement any kind of delay-based PUF. The core of an RO-PUF is a set of ring oscillators (ROs). Figure 11.1 shows the circuit diagram of an RO consisting of a single NAND gate and a number of delay elements (e.g. buffers). The delay elements are just passing on the signal. Sometimes also an odd number of inverters is used as delay elements. When the *enable* input is set to 1, the RO starts oscillating; i.e. a constant alternation between 1 and 0 is observable at the output. The oscillation frequency depends on the delay of the circular path, which is different for each device.

Figure 11.2 shows the diagram of a classical RO-PUF circuit [19]. The *challenge* selects two ROs for comparison and passes their oscillating signals to the counters. The counterfed by the faster RO eventually holds the greater value and the comparator returns 1 or 0 as the PUF response for this particular challenge. If the relative differences between the ROs are different on each device, each device has its unique signature.

As each response bit is created by comparing two ROs, there is a total of $\frac{p \cdot (p-1)}{2}$ bits. But because there are only $p!$ possibilities for p ROs to be ordered by their frequencies, the entropy of the RO-PUF's signature is in fact $log_2(p!)$ [19]. The authors of [10] introduced a post-processing procedure to generate even more signature bits from the same amount of ROs, but for simplicity we regard the standard method here. Thus, the RO-PUF belongs to the class of weak PUFs, as it does not generate an exponential amount of response bits. The term weak does not imply that the PUF is security-wise broken but rather implies that the PUF response should be concealed and impossible to be read out directly; otherwise an attacker might copy all the values and forge a device with the same response.

The remainder of this chapter is organised as follows. Section 11.2 introduces the Altera Cyclone FPGA architecture and explains how ROs are mapped to it. Section 11.3 describes in a step-by-step manner how to implement all parts of a functioning RO-PUF using the Quartus II design software and VHDL. It furthermore explains how the placement and routing of the RO circuitry must be enforced instead of leaving it to the compiler. Section 11.4 shows how to establish the communication between the FPGA and a computer via the JTAG interface. In Sect. 11.5, we share what furthermore has to be taken care of to ensure the PUF quality before Sect. 11.6 concludes the chapter.

11.2 Altera FPGA Architecture

The basic building blocks of an FPGA are its *look-up tables* (LUTs). An LUT is a circuit that can mimic any logic gate—up to a certain number of input signals—depending on its configuration. This is achieved by storing the gate's desired output value for each possible input assignment in volatile memory cells. Figure 11.3 shows an LUT with four input signals (A, B, C, D) that is configured to mimic a four-input *xor* gate. The internal architecture of an LUT is as if the input signals are connected to the select inputs of multiplexers, such that for each input assignment the appropriate stored value is passed to the LUT's output. For a four-input LUT, there are $2^4 = 16$ possible input assignments and hence output values. When these 16 output values are concatenated, they form the so-called *LUT mask*. For the configuration in Fig. 11.3, the LUT mask is 0110100110010110 or 6996 in hexadecimal.

To realise gates with less inputs, the unnecessary inputs are neglected by storing the same LUT output values regardless of whether these inputs are 0 or 1. Figure 11.4 shows two LUT configurations both realising a two-input *nand* gate. For *A nand B* (left), the same values are stored regardless of *C* and *D*. Likewise, for *C nand D* (right), *A* and *B* are disregarded.

The Altera FPGAs Cyclone I, II, III and IV provide LUTs with four inputs. In the specification of these devices you will most prominently find a figure stating the number of *logic elements* (LEs). Each LE contains a four-input LUT. There are also registers in the LEs to realise sequential circuits, but to implement delay-based PUFs the main effort is in "manually" configuring the LUTs to comprise the delay

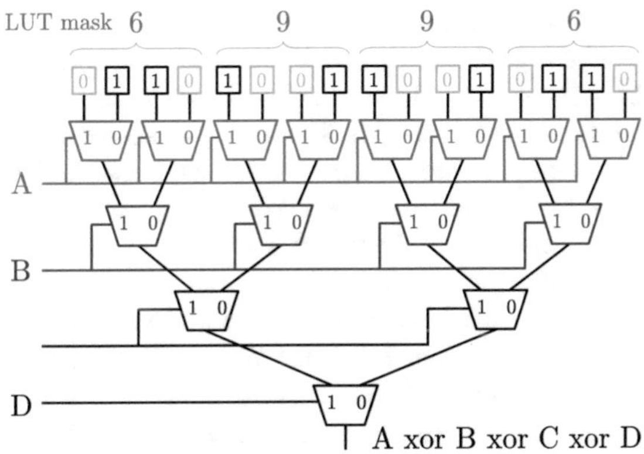

Fig. 11.3 A look-up table mimicking a 4-input *xor* gate

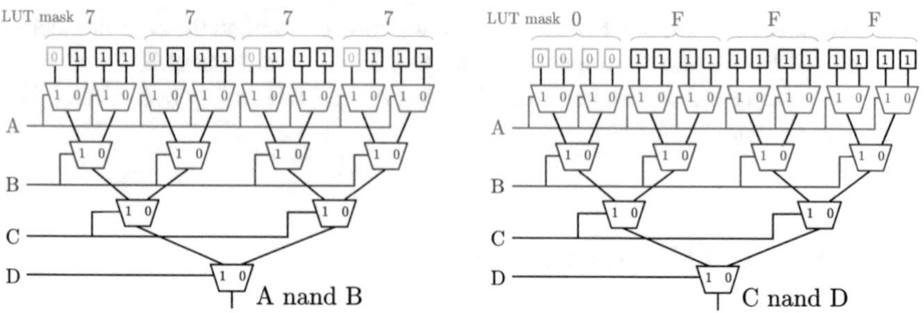

Fig. 11.4 Two different look-up table configurations, both realising a 2-input *nand* gate

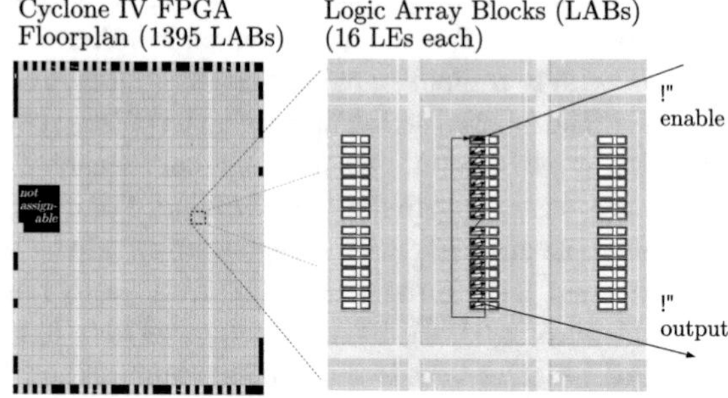

Fig. 11.5 The *left* shows the floorplan of a Cyclone IV FPGA. The *right* shows the LEs of some LABs and how an RO can be implemented in them

paths. Sequential logic is needed in the PUF evaluation logic, which can be left to the compiler. Depending on the FPGA type, a certain number of LEs is grouped in a collective called *logic array block* (LAB). On the Cyclone I, one LAB comprises 10 LEs; on the Cyclone II, III and IV it comprises 16 LEs. The Cyclone V has a slightly different architecture, in which the basic building blocks are called *adaptive logic modules* (ALMs). The Cyclone V's ALMs hold several LUTs that can be used as either two four-input LUTs or combined as one eight-input LUT. For the remainder of this chapter, our running example will assume an implementation on a Cyclone I–IV architecture. The used methods, however, can easily be adapted for a Cyclone V implementation as well.

Figure 11.5 (left) shows the floor plan of a Cyclone IV FPGA as viewed in the *Chip Planner* of Altera's Quartus II design software. Each little rectangle in the full view stands for one LAB. The right shows a zoomed-in view in which the single LEs of some LABs are visible. The routing between the LEs within an LAB is shorter than the routing from one LAB to the other. This has severe implications for the implementation of delay-based PUFs, because to elicit the subtle device-specific delay differences, delay differences shared by *all* devices must be avoided. In [5] we showed that ideal PUF uniqueness and reliability is achieved when all LEs of an LAB are used to build one RO. The arrows in Fig. 11.5 (right) show how the topmost LUT implements the RO's *nand* gate (cf. Fig. 11.1) and the remaining LUTs implement the delay elements. The next section we show this is put into practice.

11.3 Implementing the PUF

This section details how to use the Altera Quartus II software to design a delay-based PUF. Our running example is the implementation of a basic RO-PUF as described in the previous sections. The purpose of Quartus II is to make the task of creating an FPGA design easier for the designer by automatising the mapping, placement and routing to the FPGA hardware. When fine-tuning a delay-based PUF, however, it is necessary to prevent or undo some of these automatisms.

The currently newest version of Quartus II is 15.0 but the instructions of this chapter are applicable for any version. Notice that some employed functionalities—e.g. *Feature Incremental Compilation* or *LogicLock Regions*—are only available in the Quartus II Subscription Edition, not in the free Quartus II Web Edition. For academic research, a license for the Subscription Edition can be acquired without charge through the Altera University Program. For this running example, we assume that the name of the Quartus project is `myPUF` and that a Cyclone IV EP4CE22F17C6 on a Terasic DE0 Nano development board is used.

11.3.1 Defining the Hardware Components

The hardware components of our example are mainly defined in VHDL. For the communication between the FPGA and a PC we shall use the predefined *Virtual JTAG* component provided by Quartus (see Sect. 11.4), as for the n-bit multiplexers we also use one of Quartus' *Megafunctions*. All these components are then combined in a *Block Diagram/Schematic File*. Starting bottom-up though, we first define the ROs themselves.

When Quartus compiles a design, an analysis for logical redundancies is made and nodes just passing on a value like the RO's delay elements (cf. Fig. 11.1) are removed automatically. To prevent the compiler from "optimising away" the delay elements, they have to be defined as low-level primitives called LCELLs. Listing 11.1 shows the VHDL code for an RO. Line 5 sets the amount of delay elements (including the RO's nand gate) as a generic variable. Line 6 defines the input and output as Lines 10 and 11 define the internal signals. In Lines 12 and 13, the LCELL component is stated to be used in Lines 17 and 19, where the first delay element after the nand gate (`lc_0`) and the remaining delay elements (`lc_i`) are defined.

Listing 11.2 shows how an arbitrary number of ROs (given as a generic variable in Line 5) is defined reusing the RO definition (Lines 10–13). The `ros` component has one output for each of its ROs (Lines 6 and 18). In this example, the number of ROs is 16. For simplicity, all ROs are permanently activated (Line 17). In a more elaborate implementation, there could be individual input ports to activate only the desired ROs. Activating all ROs, even though only two are sampled at a time, is a waste of energy but might also be considered a countermeasure against side-channel attacks that extract RO frequencies through electromagnetic emissions [12]. Because when all ROs oscillate, it is more difficult to single out individual RO frequencies.

Listing 11.1 ro.vhd

```
LIBRARY IEEE;
USE IEEE.STD_LOGIC_1164.ALL;

ENTITY ro IS
  GENERIC (length : INTEGER := 16);
  PORT (enable : IN STD_LOGIC; output : OUT STD_LOGIC);
END ro;

ARCHITECTURE arch OF ro IS
  SIGNAL path   : STD_LOGIC_VECTOR(length DOWNTO 1);
  SIGNAL nand_out : STD_LOGIC;
  COMPONENT LCELL PORT (a_in : IN STD_LOGIC; a_out : OUT STD_LOGIC);
  END COMPONENT;
BEGIN
  nand_out <= enable nand path(length);
  output <= path(length);
  lc_0 : LCELL PORT MAP (a_in => nand_out, a_out => path(1));
  lc_gen : FOR i IN 1 TO (length-1) GENERATE
    lc_i : LCELL PORT MAP (a_in => path(i), a_out => path(i+1));
  END GENERATE;
END arch;
```

Listing 11.2 ros.vhd

```
LIBRARY IEEE;
USE IEEE.STD_LOGIC_1164.ALL;

ENTITY ros IS
  GENERIC (ros_number : INTEGER := 16; ros_length : INTEGER := 16);
  PORT (ros_out : OUT STD_LOGIC_VECTOR(ros_number-1 downto 0));
END ros;

ARCHITECTURE arch OF ros IS
  COMPONENT ro
    GENERIC (length : INTEGER);
    PORT (enable : IN STD_LOGIC; output : OUT STD_LOGIC);
  END component;
BEGIN
  ro_gen: FOR i IN 0 TO ros_number-1 GENERATE
    ro_i: ro GENERIC MAP (length => ros_length) PORT MAP (
      enable => '1',
      output => ros_out(i)
    );
  END GENERATE;
END arch;
```

After creating a new *Block Diagram/Schematic File* in Quartus (e.g. `myPUF.bdf`), a *Symbol File* for `ros.vhd` can be created by right-clicking it in the *Project Navigator* window and selecting "Create Symbol Files for Current File". Afterwards, the block symbol of `ros` can be selected with the *Symbol Tool* in the view of `myPUF.bdf`. Figure 11.6 shows the complete `myPUF.bdf` file. So far, we have only placed the ROs. Next come the 16-bit multiplexers

Listing 11.3 counter.vhd

```
1  LIBRARY IEEE;
2  USE IEEE.STD_LOGIC_1164.ALL;
3  USE IEEE.STD_LOGIC_UNSIGNED.ALL;
4
5  ENTITY counter IS
6    PORT (
7      osc    : IN STD_LOGIC;
8      run    : IN STD_LOGIC;
9      reset  : IN STD_LOGIC;
10     result : OUT STD_LOGIC_VECTOR(63 DOWNTO 0)
11   );
12 END ENTITY counter;
13
14 ARCHITECTURE arch OF counter IS
15   SIGNAL count : STD_LOGIC_VECTOR(63 DOWNTO 0);
16 BEGIN
17   result <= count;
18
19   PROCESS(osc, run, reset)
20   BEGIN
21     IF reset = '1' THEN count <= (OTHERS => '0');
22     ELSIF run = '1' THEN
23       IF RISING_EDGE(osc) THEN count <= count + 1;
24       END IF;
25     END IF;
26   END PROCESS;
27 END ARCHITECTURE arch;
```

(MUXs) (one bit for each RO), which can be found in the *Symbol Tool* under `.../quartus/libraries/megafunctions/gates/mux`.

To count the oscillations of the ROs, simple sequential counters can be used as defined in Listing 11.3. The counter value is stored in an internal 64-bit register (Line 15), that is connected to the counter's output `result` (Line 10 and 17). The input signal `osc` is connected to the output of an RO selected by a MUX. At each rising edge of `osc`, the counter value is incremented by one (Line 23), given that the `run` input signal is set to 1 (Line 22). By setting the `reset` input signal to 1, the counter value is reset to zero (Line 15).

Here, a counter width of 64 bits suffices. Depending on how long the RO oscillations are counted and on how fast they oscillate, a smaller or greater number of bits is required. Notice that for very fast ROs, a sequential counter might not be able to capture all rising edges. For such high-frequency RO-PUFs, we recommend using ripple counters instead.

Each RO-PUF response bit is created by comparing the results of two counters with a comparator logic as defined in Listing 11.4. Block symbols for `counter` and `compare` can be generated the same way as described above for `ros`. After connecting the MUXs' outputs to the counters' inputs and the counters' outputs to the comparator's input (cf. Fig. 11.6), the remaining unconnected signals are the select inputs of the MUXs, the `run` and `reset` inputs of the counters, and the `result`

Listing 11.4 compare.vhd

```
LIBRARY IEEE;
USE IEEE.STD_LOGIC_1164.ALL;

ENTITY compare IS
  PORT (
    a      : IN STD_LOGIC_VECTOR(63 DOWNTO 0);
    b      : IN STD_LOGIC_VECTOR(63 DOWNTO 0);
    result : OUT STD_LOGIC
  );
END ENTITY;

ARCHITECTURE arch OF compare IS
BEGIN
  result <= '1' WHEN a > b ELSE '0';
END arch;
```

output of the comparator. These are the signals that we want to access from outside the FPGA via the JTAG interface. Details about the communication between the FPGA and a PC are given in Sect. 11.4 together with a description of the `controller` logic and its VHDL code (Listing 11.7). The `controller` is connected to a *Virtual JTAG* component that can be generated in Quartus using the *MegaWizard Plug-In Manager*. Chose "Create a new custom megafunction variation" and then `Installed Plug-Ins/JTAG-accessible Extensions/Virtual JTAG`. As file type and filename chose VHDL and `vjtag`. Next, the *instruction register width* can be defined. As we do not need more than eight different instructions, three bits are enough. Before finishing the generation of the *Virtual JTAG* module, check that the *symbol file* `vjtag.bsf` is automatically created. Otherwise, it can also be generated afterwards like we have generated symbols from VHDL files before and like we are generating one from `controller.vhd`. It is only necessary to manually add the newly created `vjtag.vhd` file to the project files in the *Project Navigator*. After placing and connecting all block symbols correctly, the complete design in `myPUF.bdf` looks like the one in Fig. 11.6.

This design can be compiled already. However, there will be several compiler warnings like "Found combinational loop". As this is exactly what was intended with the ROs, though, these messages may be suppressed for future compiler runs. Even a first test of the PUF is possible. Among the generated signature bit values there should be an almost equal amount of 1s and 0s. An examination of several FPGAs, however, will show that many signature bits are the same for all FPGAs, which is bad PUF uniqueness.

The reason for this can be identified in the Quartus *Chip Planner*, where the mapping of the design components to the FPGA hardware is visualised. For now, the compiler was free to decide how to place the components, which is in most cases not ideal for the quality of delay-based PUFs. The following sections show how to enforce the placement and routing such that the PUF quality is greatly improved.

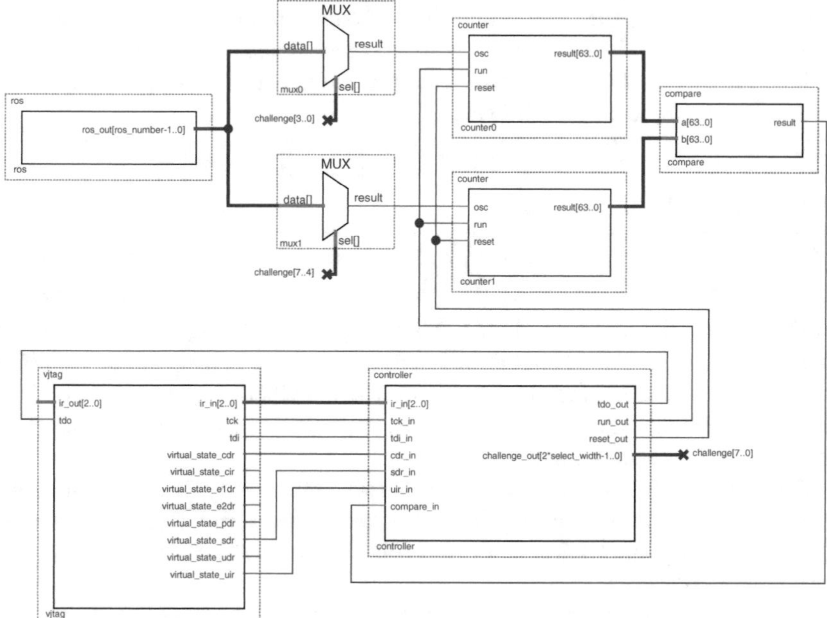

Fig. 11.6 The Quartus block diagram of the complete design

11.3.2 *Defining the LUT Placement*

One LCELL is always synthesised in one *logic element* (LE) (cf. Fig. 11.6). The Quartus compiler in normal operation places the LEs as it deems best according to timing and power supply considerations. This, however, can lead to the delay of one wire being smaller than that of another wire on *all* devices. For good uniqueness properties of delay-based PUFs it is necessary that the delays between sampled components (e.g. ROs) are as homogeneous as possible. To fine-tune a delay-based PUF, it is necessary to take control of this placement; at least for the relevant PUF components. This can be achieved in the following two ways, both of which are suited for different scenarios. Method 1 is practical for large-scale customisations of many LCELLs at once. However, a greater one-time effort is necessary to write a script producing the required file entries. Method 2 is practical for punctual customisations via graphical user interface but therefore very arduous for large amounts of LCELLs.

Method 1: Customising the Quartus Settings File

One way to define the placement is by custom entries in the *Quartus Settings File*. A great advantage of this method is that it can be achieved with the free Quartus Web Edition. It is most practical, when a script is used to generate the custom entries automatically. The *Quartus Settings File* with the `.qsf` extension is found in the Quartus project's root directory. Here, the location of each individual LCELL can be defined. To do so, the "Full Name" of the "Node" must be known by which the respective LCELL is identified in Quartus. It can be found under "Node Properties" in the Quartus *Chip Planner* when clicking on the LE of an LCELL. In the case of our running example, those "Full Names" would be, e.g.

```
|myPUF|ros:ros|ro:\ro_gen:9:ro_i|lc_0
```

for the first LCELL of RO 9, and

```
|myPUF|ros:ros|ro:\ro_gen:9:ro_i|\lc_gen:1:lc_i
|myPUF|ros:ros|ro:\ro_gen:9:ro_i|\lc_gen:2:lc_i
...
```

for the second and third LCELL of RO 9, and so forth. Notice, that `myPUF` corresponds to the name of the Quartus project as `ros` and `ro` correspond to the VHDL entities described in Listings 11.1 and 11.2. The identifiers `ro_gen` and `ro_i` are the ones used in the `GENERATE` command of `ros`, as `lc_gen` and `lc_i` are those from `ro` (cf. Listings 11.1 and 11.2).

Thus, to place for example all LCELLS of RO 9 into the LAB with the floorplan coordinates (26, 5), the following lines are added to the qsf file. The expressions `N0` to `N30` designate, which LE of that LAB should be used; `N0` is the topmost LE, `N30` the bottommost.

```
set_location_assignment LCCOMB_X35_Y31_N0  -to "ros:ros|ro:\\ro_gen:9:ro_i|lc_0"
set_location_assignment LCCOMB_X35_Y31_N2  -to "ros:ros|ro:\\ro_gen:9:ro_i|\\lc_gen:1:lc_i"
set_location_assignment LCCOMB_X35_Y31_N4  -to "ros:ros|ro:\\ro_gen:9:ro_i|\\lc_gen:2:lc_i"
set_location_assignment LCCOMB_X35_Y31_N6  -to "ros:ros|ro:\\ro_gen:9:ro_i|\\lc_gen:3:lc_i"
set_location_assignment LCCOMB_X35_Y31_N8  -to "ros:ros|ro:\\ro_gen:9:ro_i|\\lc_gen:4:lc_i"
set_location_assignment LCCOMB_X35_Y31_N10 -to "ros:ros|ro:\\ro_gen:9:ro_i|\\lc_gen:5:lc_i"
set_location_assignment LCCOMB_X35_Y31_N12 -to "ros:ros|ro:\\ro_gen:9:ro_i|\\lc_gen:6:lc_i"
set_location_assignment LCCOMB_X35_Y31_N14 -to "ros:ros|ro:\\ro_gen:9:ro_i|\\lc_gen:7:lc_i"
set_location_assignment LCCOMB_X35_Y31_N16 -to "ros:ros|ro:\\ro_gen:9:ro_i|\\lc_gen:8:lc_i"
set_location_assignment LCCOMB_X35_Y31_N18 -to "ros:ros|ro:\\ro_gen:9:ro_i|\\lc_gen:9:lc_i"
set_location_assignment LCCOMB_X35_Y31_N20 -to "ros:ros|ro:\\ro_gen:9:ro_i|\\lc_gen:10:lc_i"
set_location_assignment LCCOMB_X35_Y31_N22 -to "ros:ros|ro:\\ro_gen:9:ro_i|\\lc_gen:11:lc_i"
set_location_assignment LCCOMB_X35_Y31_N24 -to "ros:ros|ro:\\ro_gen:9:ro_i|\\lc_gen:12:lc_i"
set_location_assignment LCCOMB_X35_Y31_N26 -to "ros:ros|ro:\\ro_gen:9:ro_i|\\lc_gen:13:lc_i"
set_location_assignment LCCOMB_X35_Y31_N28 -to "ros:ros|ro:\\ro_gen:9:ro_i|\\lc_gen:14:lc_i"
set_location_assignment LCCOMB_X35_Y31_N30 -to "ros:ros|ro:\\ro_gen:9:ro_i|\\lc_gen:15:lc_i"
```

Method 2: Using Design Partitions and LogicLock Regions

Another way to enforce the placement of LCELLs into specific LEs makes use of the *LogicLock* feature of the licensed Quartus II Subscription Edition. The advantage of this method is that the LCELLs can be easily relocated on the FPGA's floorplan via "drag-and-drop" in the Quartus *Chip Planner*. This manual procedure, is thus only advisable for moderate amounts of LCELLs.

First, a *Design Partition* for each entity of the delay-based PUF (here for each RO) has to be created. This can be done in the *Chip Planner's* "Design Partitions Window". Afterwards, a LogicLock Region (LLR) for each Design Partition must be created by right-clicking on the newly created partition entries and selecting "LogicLock Region" → "Create New LogicLock Region". The newly created LLRs are then shown in the *Chip Planner's* "LogicLock Region Window". For the LLR of each RO, set "Size: fixed" and "State: Locked", allowing to specify their width, height and position. Setting "Reserved: On" determines that no other logic than the corresponding *Design Partition* is placed in the respective LLR. Then go back to the

Quartus main window and compile the whole project once more. Afterwards, the *Chip Planner* shows the ROs placed in the defined LLR locations.

However, the placement of the LCELLs within the LEs of an LLR is still determined automatically by the compiler. We may now use the mouse cursor to drag-and-drop the LCELLs from one LE to another, until we have the desired configuration. For each drag-and-drop operation, a "change" is added to the *Chip Planner's* "Change Manager" window. This window is also where the button "Check and Save All Netlist Changes" is found. When all drag-and-drop changes have been made, click this button to start another partial compilation that relocates the LCELLs.

A new complete compiler run, however, would always undo these manual changes and one has to go back to the "Change Manager" to reapply them each time. Furthermore, it is not possible to relocate the LLRs keeping the custom LE placement within them. To keep it, one has to go through the following process. First, in the Quartus main window select "Assignments" → "Back-Annotate Assignments" → "Pin, cell & device assignments". This saves the current placement of the LCELLs such that they are not undone when the project is compiled again. Then perform another full compiler run and go back to the *Chip Planner's* "Design Partitions Window" and for the *Design Partition* of each RO, set "Netlist Type: Post-Fit" and "Fitter Preservation Level = Placement". This forces future compiler runs to use the last placement within these *Design Partitions*. Thus, it is now possible to relocate the LLRs keeping their internal LCELL placement.

When the LLRs are moved to new locations, the previously back-annotated assignments are no longer valid and should be removed. This is done in the Quartus main window by selecting "Assignments" → "Remove Assignments" → "Pin, Location & Routing Assignments". Notice though, that this also removes all previous pin assignments; like, e.g. which pins of the FPGA are connected to an external clock, LEDs or push buttons. These have to be redefined in the Quartus Pin Planner. Ideally, all necessary assignments have been exported to a file from which they can easily be imported again.

11.3.3 Defining the LUT Routing

The previous section showed how to enforce the placement of design components. With the described methods, it is possible to specify in which logic elements (LEs) the single components are implemented. But there is no immediate way to specify which of the four LUT inputs of an LCELL is used for which input signal. Leaving this up to the compiler may also lead to inferior uniqueness as outlined in the following.

Consider the simple example of an LUT that only passes on its one input signal to its output—a typical application in delay-based PUFs on FPGAs. To use LUT input A, the LUT mask is set to $1010101010101010 = AAAA$ (cf. Fig. 11.3); for LUT input B to $1100110011001100 = CCCC$; for LUT input C to $1111000011110000 = F0F0$; and for LUT input D to $1111111100000000 = FF00$. All of these configurations logically perform the same task. A glance at Fig. 11.3, however, reveals that the path from input A to the output is the longest of all the inputs, whereas the path from input D to the output is the shortest. In the real FPGA hardware it might not necessarily be the case that the delay of input D to the output is indeed the fastest. This depends on how the hardware is really structured. But there are definitely significant delay differences between the different LUT inputs.

For the implementation of a delay-based PUF, such differences can have a severe impact on the PUF quality. If, for example the LCELLs of one RO x mainly use their LUT input A and the LCELLs of another RO y mainly use their input D, RO x will most likely be slower than RO y on all FPGAs; leading to poor uniqueness. Such scenarios do in fact occur, if the LUT routing is left to the compiler. This section does therefore presents how to enforce the LUT routing as well.

To manually change the used LUT inputs, one has to open the Quartus *Resource Property Editor* by double-clicking on an LE in the *Chip Planner*. There, the "Signal Name" of each LUT input can be edited or removed. To change, e.g. a LUT using input C to using input D, copy the signal name for input C and paste it for input D. Then remove the connection for input C. Lastly, change the LUT mask from $F0F0$ to $FF00$. Afterwards, the corresponding changes are listed in the *Chip Planner's* "Change Manager", and can be applied by clicking the "Check and Save All Netlist Changes" button.

Instead of applying all changes manually, which is rather tedious and error-prone for larger amounts of LUTs, it is advisable to write a script to perform them automatically. With a right-click in the "Change Manager", any changes can be exported to a TCL file (e.g. `changes.tcl`) that performs the changes when called with the console command:

```
quartus_cdb -t changes.tcl
```

Thus, applying the above three example changes to, e.g. LCELL 3 of RO 0 can be performed by the TCL script given in Listing 11.5. The actual changes are defined in Lines 19–21 (add the connection to LUT input D), Lines 43–44 (remove the connection of LUT input C), and Lines 63–65 (change the LUT mask). But the preceding `set node_properties` commands (Lines 9–18, 30–42, 53–62) are necessary, too. Here, the state of the LCELL prior to the change has to be given. Notice that the "fanins" for the first change (Lines 13–17) only include the one connection to LUT input C coming from the previous LCELL 2 (`lc_gen:2`). After adding the connection to LUT input D, the "fanins" of the next change must also include that connection (Lines 34–41). After removing the connection to LUT input C, the "fanins" of the third change is only left with the connection to LUT input D (Lines 58–60). Apart from the "fanins", the current LUT mask has to be stated in each `set node_properties` command (Lines 12, 33, 56). Would there be more changes to this LCELL in this TCL script, the new LUT mask $FF00$ would have to be stated for all subsequent changes.

Listing 11.5 changes.tcl

```
package require ::quartus::chip_planner
package require ::quartus::project
load_chip_planner_utility_commands
project_open myPUF -revision myPUF
read_netlist
set had_failure 0

# Adding LUT input D of LCELL 3 of RO 0.
set node_properties [ node_properties_record #auto \
  -node_name |myPUF|ros:ros|ro:\\ro_gen:0:ro_i|\\lc_gen:3:lc_i \
  -node_type LCCOMB_CII -op_mode normal -data_to_lut_c "Data C" \
  -sum_lut_mask F0F0 \
  -fanins [ list \
    [ fanin_record #auto -dst {-port_type DATAC -lit_index 0} \
      -src {-node_name |myPUF|ros:ros|ro:\\ro_gen:0:ro_i|\\lc_gen:2:lc_i \
      -port_type COMBOUT -lit_index 0}  -delay_chain_setting -1 ] \
  ] \
]
set result [ make_ape_connection_wrapper \
  $node_properties |myPUF|ros:ros|ro:\\ro_gen:0:ro_i|\\lc_gen:3:lc_i DATAD 0 \
  |myPUF|ros:ros|ro:\\ro_gen:0:ro_i|\\lc_gen:2:lc_i COMBOUT 0 -1 ]
if { $result == 0 } {
  set had_failure 1
  puts "Use the following information to evaluate how to apply this change."
  dump_node $node_properties
}
remove_all_record_instances

# Removing LUT input C of LCELL 3 of RO 0.
set node_properties [ node_properties_record #auto \
  -node_name |myPUF|ros:ros|ro:\\ro_gen:0:ro_i|\\lc_gen:3:lc_i \
  -node_type LCCOMB_CII  -op_mode normal -data_to_lut_c "Data C" \
  -sum_lut_mask F0F0 \
  -fanins [ list \
    [ fanin_record #auto -dst {-port_type DATAC -lit_index 0} \
      -src {-node_name |myPUF|ros:ros|ro:\\ro_gen:0:ro_i|\\lc_gen:2:lc_i \
        -port_type COMBOUT -lit_index 0}  -delay_chain_setting -1 ] \
    [ fanin_record #auto -dst {-port_type DATAD -lit_index 0} \
      -src {-node_name |myPUF|ros:ros|ro:\\ro_gen:0:ro_i|\\lc_gen:2:lc_i \
        -port_type COMBOUT -lit_index 0}  -delay_chain_setting -1 ] \
  ] \
]
set result [ remove_ape_connection_wrapper \
  $node_properties |myPUF|ros:ros|ro:\\ro_gen:0:ro_i|\\lc_gen:3:lc_i DATAC 0 ]
if { $result == 0 } {
  set had_failure 1
  puts "Use the following information to evaluate how to apply this change."
  dump_node $node_properties
}
remove_all_record_instances

# Changing LUT Mask of LCELL 3 of RO 0.
set node_properties [ node_properties_record #auto \
  -node_name |myPUF|ros:ros|ro:\\ro_gen:0:ro_i|\\lc_gen:3:lc_i \
  -node_type LCCOMB_CII -op_mode normal -data_to_lut_c "Data C" \
  -sum_lut_mask F0F0 \
  -fanins [ list \
    [ fanin_record #auto -dst {-port_type DATAD -lit_index 0} \
      -src {-node_name |myPUF|ros:ros|ro:\\ro_gen:0:ro_i|\\lc_gen:2:lc_i \
        -port_type COMBOUT -lit_index 0}  -delay_chain_setting -1 ] \
  ] \
]
set result [ set_lutmask_wrapper \
  $node_properties |myPUF|ros:ros|ro:\\ro_gen:0:ro_i|\\lc_gen:3:lc_i \
  "Sum LUT Mask" FF00 ]
if { $result == 0 } {
  set had_failure 1
  puts "Use the following information to evaluate how to apply this change."
  dump_node $node_properties
}
remove_all_record_instances

# Apply the changes.
set drc_result [check_netlist_and_save]
if { $drc_result == 1 } {
  puts "check_netlist_and_save: SUCCESS"
} else {
  puts "check_netlist_and_save: FAIL"
}
if { $had_failure == 1 } {
  puts "Not all set operations were successful"
}
project_close
```

```
1  ...
2  signal_name = ros:ros|ro:\ro_gen:0:ro_i|\lc_gen:2:lc_i {
3    LOCAL_LINE:X34Y33S0I2;
4    dest = ( ros:ros|ro:\ro_gen:0:ro_i|\lc_gen:3:lc_i, DATAA ), route_port = DATAC;
5  }
6  signal_name = ros:ros|ro:\ro_gen:0:ro_i|\lc_gen:3:lc_i {
7    LOCAL_LINE:X34Y33S0I3;
8    dest = ( ros:ros|ro:\ro_gen:0:ro_i|\lc_gen:4:lc_i, DATAA ), route_port = DATAD;
9  }
10 signal_name = ros:ros|ro:\ro_gen:0:ro_i|\lc_gen:4:lc_i {
11   LOCAL_LINE:X34Y33S0I4;
12   dest = ( ros:ros|ro:\ro_gen:0:ro_i|\lc_gen:5:lc_i, DATAA ), route_port = DATAC;
13 }
14 ...
```

Thus, a script generating the TCL file must—for each change—take into account all previous changes made to the respective LCELL, which raises the question how the initial states of all LCELLs can be known. As the compiler may chose the LUT inputs arbitrarily, there is no way to predict the routing. Instead, it has to be extracted after the compilation using the "Back-Annotate Assignments" feature already applied in Method 2 of Sect. 11.3.2. Only here, "Assignments" → "Back-Annotate Assignments" → "Pin, cell, routing & device assignments" is selected, as opposed to just "Pin, cell & device assignments". This results in the generation of the *Routing Constraints File* `myPUF.rcf` in the project's main directory. Within this file there are—albeit not in this correct order—entries of the following form:

To extract which LUT inputs are used for which LCELL, one has to look at the `dest` entry of the signal connected to that input. Thus, Lines 2–4 indicate that the output of LCELL 2 is connected to LUT input C (`DATAC`) of LCELL 3; just as Lines 6–8 indicate that the output of LCELL 3 is connected to LUT input D (`DATAD`) of LCELL 4. The LUT masks are not explicitly stated in the *Routing Constraints File*, but knowing the LUT inputs allows for them to be inferred.

The complete workflow for defining the LUT routing is hence:

1. Compile the Quartus project enforcing the desired LUT placement as explained in Sect. 11.3.2.
2. Use the "Back-Annotate Assignments" feature to generate the *Routing Constraints File* (`.rcf`).
3. Check the `.rcf` file to extract original LUT inputs used by the compiler.
4. Create a TCL script `changes.tcl` to change the LUT inputs as desired.
5. Run `quartus_cdb -t changes.tcl` to execute the TCL script and apply the changes.

11.4 Communication Between PC and FPGA

When the Quartus project is compiled successfully, the PUF implementation is ready to be tested on real FPGAs. This section describes how the communication between an FPGA and a PC can be achieved via the FPGA's JTAG interface. Most FPGA circuit boards have a USB port through which the JTAG interface can be accessed and Quartus provides several TCL commands for all necessary communication tasks.

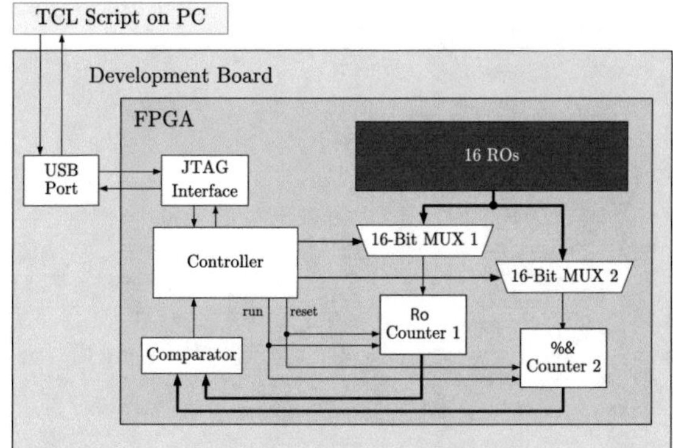

Fig. 11.7 A schematic view of the communication between the TCL script running on a PC and the hardware controller on the FPGA

Figure 11.7 shows a schematic view of the communication flow. Basically, the TCL script tells the controller which two ROs should be compared to generate a single response bit and afterwards reads out this bit.

For our running example, let the TCL script including these communication commands be `myPUF_test.tcl`, given in Listing 11.6. In order to use the specific TCL commands it is necessary to run the script within Quartus SignalTap, which is done by entering the following in the command line:

```
quartus_stp -t myPUF_test.tcl
```

We will go through `myPUF_test.tcl` step by step and at the same time explain its counterpart on the FPGA: i.e. the controller already mentioned in the end of Sect. 11.3.1. Its VHDL code `controller.vhd` is given in Listing 11.7 and Fig. 11.6 showed how is integrated in the overall design.

`myPUF_test.tcl` is divided into two parts: Lines 1–22 define the function `get_response` that is called in Lines 34–36 of the script's main section, each time to sample one PUF response bit from the FPGA. Its argument `challenge` is an eight-bit string, whose first four bits define one RO and last four bits the other RO to be compared. Four bits per RO are sufficient here as we have only 16 ROs.

Line 24 executes `quartus_pgm` to programme the compiled bitstream onto the FPGA. Its `--cable` argument selects the USB cable to be used. If there is only one, the argument is 1 as the counting of cables starts at 1. The `-o` argument states the path of the compiled `.sof` bitstream file; appended to the prefix "`p;`".

Lines 26–29 select the USB cable and FPGA, needed later when opening the FPGA in Line 31. Line 26 gets a list of all FPGA USB cables connected to the PC. Line 27 selects one of those and stores its reference in the variable `hardware_name`. Notice that the list index here starts at 0. This `hardware_name` is used in Line 28 to get a list of FPGA devices found at this cable. Normally, there is only one FPGA per cable such that the first and only list element 0 is selected in Line 29 and stored in the variable `device_name`. Both variables, `hardware_name` and `device_name` are used as arguments of `open_device` in Line 31 to start communicating with the FPGA. The locking of the device in

Listing 11.6 myPUF_test.tcl

```
1 proc get_response {challenge} {
2   set PUSH_CHALLENGE 0
3   set POP_RESPONSE   1
4   set START_COUNTERS 2
5   set STOP_COUNTERS  3
6   set RESET_COUNTERS 4
7
8   device_virtual_ir_shift -instance_index 0 -ir_value $PUSH_CHALLENGE
9   device_virtual_dr_shift -instance_index 0 \
10    -length [string length $challenge] -dr_value $challenge
11
12  device_virtual_ir_shift -instance_index 0 -ir_value $STOP_COUNTERS
13  device_virtual_ir_shift -instance_index 0 -ir_value $RESET_COUNTERS
14  device_virtual_ir_shift -instance_index 0 -ir_value $START_COUNTERS
15  after 20
16  device_virtual_ir_shift -instance_index 0 -ir_value $STOP_COUNTERS
17  device_virtual_ir_shift -instance_index 0 -ir_value $POP_RESPONSE
18  set response [device_virtual_dr_shift -instance_index 0 \
19    -length 1 -dr_value 0]
20
21  puts "$challenge: $response"
22 }
23
24 exec quartus_pgm --cable=1 --mode=JTAG -o "p;myPUF/output_files/myPUF.sof"
25
26 set hardware_names [get_hardware_names]
27 set hardware_name [lindex $hardware_names 0]
28 set device_names [get_device_names -hardware_name $hardware_name]
29 set device_name [lindex $device_names 0]
30
31 open_device -device_name $device_name -hardware_name $hardware_name
32 device_lock -timeout 10000
33
34 get_response "00000001"
35 get_response "00010010"
36 get_response "00100011"
37
38 device_unlock
39 close_device
```

Line 32 is necessary for the communication commands in the `get_response` function. In the end (Lines 38 and 39), the device is unlocked and closed again.

Within the `get_response` function, the first commands (Lines 2–6) define custom instructions used in the communication with the FPGA. Each instruction is really an integer to be transmitted via JTAG to the FPGA's *instruction register* with the `device_virtual_ir_shift` command (Lines 8, 12, 13, 14, 16 and 17). The `-instance_index` argument must be set even if there is only one *Virtual JTAG* instance in our implementation. The `-ir_value` argument is the instruction to be transmitted. We will explain the purpose each instruction while going through the remainder of `get_response`. Simultaneously, we will go through the code of `controller.vhd` (Listing 11.7) to explain the interaction between the TCL script and the FPGA.

The controller's input ports `ir_in`, `tck_in`, `tdi_in`, `cdr_in`, `sdr_in` and `uir_in` are connected to the outputs of the *Virtual JTAG* component as seen

Listing 11.7 controller.vhd

```
LIBRARY IEEE;
USE IEEE.STD_LOGIC_1164.ALL;

ENTITY controller IS
  GENERIC (select_width : INTEGER := 4);
  PORT (
    ir_in          : IN STD_LOGIC_VECTOR(2 DOWNTO 0);
    tck_in         : IN STD_LOGIC;
    tdi_in         : IN STD_LOGIC;
    cdr_in         : IN STD_LOGIC;
    sdr_in         : IN STD_LOGIC;
    uir_in         : IN STD_LOGIC;
    compare_in     : IN STD_LOGIC;

    tdo_out        : OUT STD_LOGIC;
    run_out        : OUT STD_LOGIC;
    reset_out      : OUT STD_LOGIC;
    challenge_out  : OUT STD_LOGIC_VECTOR(2*select_width-1 DOWNTO 0)
  );
END controller;

ARCHITECTURE arch OF controller IS
  CONSTANT PUSH_CHALLENGE : STD_LOGIC_VECTOR(2 DOWNTO 0) := "000";
  CONSTANT POP_RESPONSE   : STD_LOGIC_VECTOR(2 DOWNTO 0) := "001";
  CONSTANT START_COUNTERS : STD_LOGIC_VECTOR(2 DOWNTO 0) := "010";
  CONSTANT STOP_COUNTERS  : STD_LOGIC_VECTOR(2 DOWNTO 0) := "011";
  CONSTANT RESET_COUNTERS : STD_LOGIC_VECTOR(2 DOWNTO 0) := "100";
  SIGNAL challenge : STD_LOGIC_VECTOR(2*select_width-1 DOWNTO 0);
  SIGNAL run       : STD_LOGIC;
BEGIN
  challenge_out <= challenge;
  run_out       <= run;
  reset_out     <= '1' WHEN ir_in = RESET_COUNTERS and uir_in = '1'
    ELSE '0';

  PROCESS(tck_in)
  BEGIN
    IF RISING_EDGE(tck_in) THEN
      IF ir_in = PUSH_CHALLENGE and sdr_in = '1' THEN
        challenge <= tdi_in & challenge(2*select_width-1 DOWNTO 1);
      ELSIF ir_in = POP_RESPONSE and cdr_in = '1' THEN
        tdo_out <= compare_in;
      ELSIF ir_in = START_COUNTERS and uir_in = '1' THEN run <= '1';
      ELSIF ir_in = STOP_COUNTERS  and uir_in = '1' THEN run <= '0';
      END IF;
    END IF;
  END PROCESS;
END arch;
```

in Fig. 11.6. `tck_in` is the JTAG clock, synchronising the *Virtual JTAG* operations with the sequential operations of the controller (Lines 36 and 38). `ir_in` is connected to the JTAG *instruction register*, determining the state of the controller at any time (Lines 33, 39, 41, 43 and 44 of `controller.vhd`). As already mentioned, the loading of an instruction into `ir_in` from the TCL script is done by the `device_virtual_ir_shift` command. In Lines 23–27 of `controller.vhd` the instructions are defined equivalently to the definitions in the TCL script.

The JTAG communication only has a bit-width of one. Thus, a value of more bits—like the three-bit instruction—must be shifted through bit by bit. The shifting of instructions is done automatically by `device_virtual_ir_shift`. Once it is complete, the `uir_in` flag is set to 1 indicating that the most recent *update* of the *instruction register* is complete. Therefore, it is always necessary to check for *uir_in* = '*1*' when interpreting the value of `ir_in` (Lines 33, 43 and 44 of `controller.vhd`). Lines 39 and 41 are exceptions as they work with the flags `cdr_in` and `sdr_in`, which are themselves indicators that the last instruction register shift has been finished.

Thus, Line 8 of `myPUF_test.tcl` shifts the `PUSH_CHALLENGE` instruction into the *instruction register*. Afterwards, Line 9 performs a *data shift* of the eight-bit PUF challenge. The `-length` argument states the length of the `-dr_value` argument, which is thereby shifted through the `tdi_in` port; one bit per `tck_in` clock cycle. This is matched by Lines 39 and 40 of `controller.vhd`, in which one bit per cycle is shifted into the controller's `challenge` register. The `sdr_in` flag is set to 1 until the shift is completed.

With the `challenge` register being connected to the `challenge_out` port, the multiplexers now have the desired values assigned to their select inputs (cf. Figure 11.6) and the frequencies of the selected ROs can be sampled. This is done in Lines 12–16 of `myPUF_test.tcl` by loading the `STOP_COUNTERS`, `RESET_COUNTERS` and `STOP_COUNTERS` instructions, followed by a waiting time of 20 ms, followed by loading the `STOP_COUNTERS` instruction. The first `STOP_COUNTERS` in Line 12 is merely a precaution to make sure the ROs are really stopped. In `controller.vhd`, the `run` register—connected to the `run_out`—is determined by the `START_COUNTERS` and `STOP_COUNTERS` instructions (Lines 43 and 44), as the `reset_out` port is determined by the `RESET_COUNTERS` instruction (Line 33).

After the RO sampling and hence the response bit generation is finished, it must be extracted from the FPGA. This is initiated in Line 17 of `myPUF_test.tcl` by loading the `POP_RESPONSE` instruction, followed by another `device_virtual_dr_shift` in Line 18, whose result is stored in the `response` variable. In Line 21, the result is printed to the screen.

In the first clock cycle of `device_virtual_dr_shift`, the `cdr_in` flag—not yet the `sdr_in` flag—is set, signalling the controller to *capture* the value to be shifted out. This is done in Line 41 of `controller.vhd`, where `compare_in` is simply connected to the `tdo_out` port. If we had to shift out a value with more than one bit, we would have had to store the value in a register upon `cdr_in`, connect the first bit of that register to `tdo_out`, and shift it on—one bit per subsequent clock cycle—as long as `sdr_in` is still set. But as our response bit here only consists of one bit, this minimal implementation is sufficient.

11.5 Traps and Pitfalls

When designing a delay-based PUF on FPGAs, there are certain peculiarities to watch out for. When the goal is to achieve good PUF uniqueness, the delays must be so homogeneous that the impact of device-specific process variations can be large enough to "tip the scales"; i.e. to determine the PUF signature values individually for each device. Furthermore, when good PUF reliability is desired, it is necessary to prevent all kinds of delay variance while the device is in operation.

So far we have shown how to cater for PUF uniqueness by making the placement and routing of delay-relevant PUF structures as homogeneous as possible. The thereby achieved PUF quality is already within acceptable limits [5]. Further experimental results, however, show that even with homogeneous routing, there are still delay biases which are the same for all devices leading to a decreased uniqueness. Furthermore, it was shown in how far switching activity on the remaining circuit can influence the delays of the PUF circuitry. Such effects have to be considered, when good PUF quality shall be achieved.

Delay biases based on location

For reasons lying in the physical properties of an FPGA chip, the logic elements in different locations can have tendencies for faster or slower delays, called *biases*. This was first reported for Xilinx FPGAs by [11, 13]. We conducted a thorough analysis on Altera FPGAs regarding location dependent delay biases by evaluating the results of counters connected to ROs placed in all possible locations. The averages of 38 Cyclone IV and 20 Cyclone II FPGAs were sampled and visualised as shown in Fig. 11.8. It can be seen that the biases are quite different in different locations. Furthermore, depending on the FPGA technology, the bias distribution is very different as well.

If the bias difference between two locations is too large, a delay-based PUF will generate the same response for all devices when comparing the delays of these locations. This is because the device-specific delay difference caused by process variation is too subtle to overcome the bias [4]. Thus, when implementing a delay-based PUF, the compared locations must either have very equal biases—as suggested in [13] which, however, drastically reduces the amount of possible comparisons—or the biases have to be compensated otherwise. In [4] we showed that the biases can be estimated with sufficient accuracy based on just a small sample of devices. When the biases are known, they can be compensated on each device at the time the PUF response is generated.

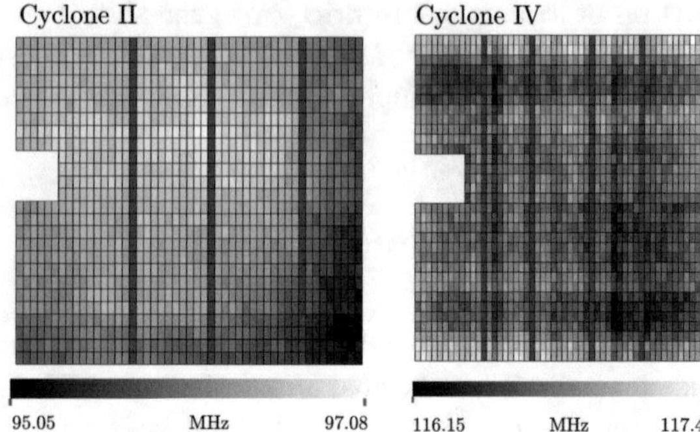

Fig. 11.8 The colour at the position of an LAB represents the average frequency of a ring oscillator implemented there. The average is calculated from sampling 38 Cyclone IV and 20 Cyclone II FPGAs

Switching activity of non-PUF circuitry

The reliability of a delay-based PUF is hampered if the delays of compared components change in opposing directions. In the case of an RO-PUF for example, there is no change in the PUF response as long as the delays of compared ROs change in the same direction: the faster RO will still be the faster. But if the delays change in different directions the PUF response will be different.

In our experiments, we found that delays are influenced by switching activity on the FPGA. There are two main effects: heating and energy consumption. It could be asserted that heating only has a global effect influencing all locations in the same way. Energy consumption, however, influences closer locations more than remote locations. The experiment shown in Fig. 11.9 illustrates this. For the first 90 min, only a single RO is running whose start frequency depends on the initial temperature of the FPGA. The oscillation of the RO is slowly heating up the chip, reducing its frequency. After about 90 min, the additional 79 ROs are activated. We see an immediate frequency drop which is larger for the design in which the sampled RO is right in the middle of the additional ROs. Afterwards, the switching activity of the additional ROs is heating up the FPGA even further. At 180 min, the additional ROs are deactivated again, whereupon the frequencies of both ROs are again the same.

Thus, when implementing a delay-based PUF, it must be ensured that there is always the same kind of switching activity in the non-PUF circuitry during the response generation. Or these temporary biases induced by switching activity must be compensated just as those induced by the physical properties described above. One might even consider using the non-PUF switching activity as another layer of security, such that the correct response is only generated if and only if the right non-PUF activity is executed. On the other hand, one must consider that an attacker might find ways to manipulate the non-PUF activity in ways to alter the PUF response of an arbitrary device.

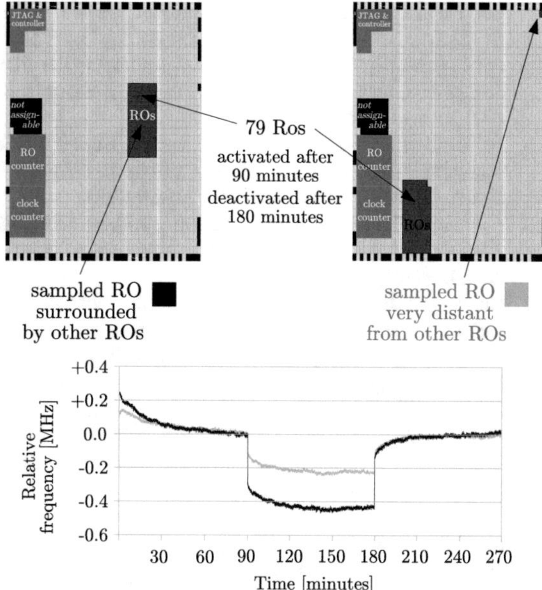

Fig. 11.9 Experiments with two different designs show that energy consumption but not heating of switching activity can influence ROs depending on their distance to the source of switching activity. The two curves in the plot show the relative frequencies of a sampled RO in both designs

11.6 Conclusion

We have shown the methods necessary to ensure homogeneous placement and routing of delay-based PUFs on Altera Cyclone FPGAs. Without these PUF uniqueness is not achievable. A complete description detailing all required steps has so far been missing in the literature. Furthermore, we have shown how the communication between the FPGA and a PC can be managed, enabling also novice readers to embark directly on their own PUF experiments.

Section 11.5 included insights from the authors' most recent research results. Future work will consist of developing and refining methods to cope for example with location based delay biases or with the biases induced by different non-PUF switching activity. The application of programmable delay lines [3] in this field is a very promising perspective.

References

1. Bohm C, Hofer M, Pribyl W. A microcontroller SRAM-PUF. In: 2011 5th international conference on network and system security (NSS); 2011. p. 269–73. doi:10.1109/ICNSS.2011.6060013.
2. Bossuet L, Ngo XT, Cherif Z, Fischer V. A PUF based on a transient effect ring oscillator and insensitive to locking phenomenon. IEEE Trans Emer Top Comput. 2014;2(1):30–6. doi:10.1109/TETC.2013.2287182.

3. Chen YY, Huang JL, Kuo T. Implementation of programmable delay lines on off-the-shelf GPGAS. In: AUTOTESTCON, IEEE; 2013. p. 1–4. doi:10.1109/AUTEST.2013.6645040.
4. Feiten L, Martin T, Sauer M, Becker B. Improving RO-PUF quality on FPGAs by incorporating design-dependent frequency biases. In: IEEE European test symposium; 2015. doi:10.1109/ETS.2015.7138749.
5. Feiten L, Spilla A, Sauer M, Schubert T, Becker B. Implementation and analysis of ring oscillator PUFs on 60 nm Altera Cyclone FPGAs. Inf Secur J Glob Perspect. 2013;22(5–6):265–73. doi:10.1080/19393555.2014.891281.
6. Fournier J, Loubet-Moundi P. Memory address scrambling revealed using fault attacks. In: 2010 workshop on fault diagnosis and tolerance in cryptography (FDTC); 2010. p. 30–6. doi:10.1109/FDTC.2010.13.
7. Guajardo J, Kumar SS, Schrijen GJ, Tuyls P. FPGA intrinsic PUFs and their use for IP protection. In: Proceedings of the 9th international workshop on cryptographic hardware and embedded systems. Springer; 2007. p. 63–80. doi:10.1007/978-3-540-74735-2_5.
8. Kumar S, Guajardo J, Maes R, Schrijen GJ, Tuyls P. The butterfly PUF: protecting ip on every fpga. In: IEEE international workshop on hardware-oriented security and trust, 2008. HOST; 2008. p. 67–70. doi:10.1109/HST.2008.4559053.
9. Lim D, Lee J, Gassend B, Suh G, van Dijk M, Devadas S. Extracting secret keys from integrated circuits. IEEE Trans Very Large Scale Integr VLSI Syst. 2005;13(10):1200–5. doi:10.1109/TVLSI.2005.859470.
10. Maiti A, Kim I, Schaumont P. A robust physical unclonable function with enhanced challenge-response set. IEEE Trans Inf Forensics Secur. 2012;7(1):333–45. doi:10.1109/TIFS.2011.2165540.
11. Maiti A, Schaumont P. Improving the quality of a physical unclonable function using configurable ring oscillators. In: International conference on field programmable logic and applications, 2009. FPL; 2009. p. 703–7. doi:10.1109/FPL.2009.5272361.
12. Merli D, Schuster D, Stumpf F, Sigl G. Semi-invasive EM attack on FPGA RO PUFs and countermeasures. In: Proceedings of the workshop on embedded systems security, WESS '11. ACM; 2011. p. 2:1–2:9. doi:10.1145/2072274.2072276.
13. Merli D, Stumpf F, Eckert C. Improving the quality of ring oscillator PUFs on FPGAs. In: Proceedings of the 5th workshop on embedded systems security; 2010. p. 9:1–9:9. doi:10.1145/1873548.1873557.
14. Note JB, Rannaud E. From the bitstream to the netlist. In: Proceedings of the 16th Int'l ACM/SIGDA symposium on FPGAs, FPGA '08. ACM; 2008. p. 264. doi:10.1145/1344671.1344729.
15. Rührmair U, Sölter J, Sehnke F. On the foundations of physical unclonable functions; 2009. https://eprint.iacr.org/2009/277.pdf.
16. Sklavos N. Securing communication devices via physical unclonable functions (PUFs). In: Reimer H, Pohlmann N, Schneider W, editors. ISSE 2013 securing electronic business processes. Fachmedien Wiesbaden: Springer; 2013. p. 253–61. doi:10.1007/978-3-658-03371-2_22.
17. Skorobogatov S. Flash memory 'bumping' attacks. In: Mangard S, Standaert FX, editors. Cryptographic hardware and embedded systems, CHES 2010. Lecture notes in computer science, vol. 6225. Berlin Heidelberg: Springer; 2010. p. 158–172. doi:10.1007/978-3-642-15031-9_11.
18. Suh G, O'Donnell C, Devadas S. Aegis: a single-chip secure processor. IEEE Des Test Comput. 2007;24(6):570–80. doi:10.1109/MDT.2007.179.
19. Suh GE, Devadas S. Physical unclonable functions for device authentication and secret key generation. In: Proceedings of the 44th annual design automation conference; 2007. p. 9–14. doi:10.1145/1278480.1278484.
20. Yin CE, Qu G. Temperature-aware cooperative ring oscillator PUF. In: Proceedings of the 2009 IEEE international workshop on hardware-oriented security and trust; 2009. p. 36–42. doi:10.1109/HST.2009.5225055.

21. Yu H, Leong P, Hinkelmann H, Moller L, Glesner M, Zipf P. Towards a unique FPGA-based identification circuit using process variations. In: International conference on field programmable logic and applications, 2009. FPL; 2009. p. 397–402. doi:10.1109/FPL.2009.5272255.
22. Yu MD, Devadas S. Secure and robust error correction for physical unclonable functions. IEEE Des Test Comput. 2010;27(1):48–65. doi:10.1109/MDT.2010.25.
23. Zhang J, Lin Y, Lyu Y, Qu G. A PUF-FSM binding scheme for FPGA IP protection and pay-per-device licensing. IEEE Trans Inf Forensics Secur. 2015;10(6):1137–50. doi:10.1109/TIFS.2015.2400413.

Chapter 12
Implementation and Analysis of Ring Oscillator Circuits on Xilinx FPGAs

Mario Barbareschi, Giorgio Di Natale and Lionel Torres

12.1 Introduction

As security of digital applications relies on trustworthy hardware platforms, new design challenges emerge from requirements of in-field applications which adopt field programmable gate arrays (FPGAs) as the hardware implementation technology. Indeed, the FPGA technology, contrary to the application-specific integrated circuits (ASICs), is able to be configured and updated in-field, out of the foundry, by means of a configuration file called bitstream. Its design methodology allows to fast prototype hardware devices and to avoid expensive nonrecurring engineering costs, which characterize ASIC projects, especially when the production scale is limited to few units. These advantages are really attractive and have created a new huge market segment around such devices.

However, as they are reconfigurable, FPGAs are more exposed to security attacks than ASICs. For instance, the intellectual property (IP) theft attack can be accomplished by read out the bitstream from the internal configuration memory or from external flash memories, once the application is deployed. Bitstream theft enables cloning of the original device into compatible devices or, by exploiting reverse engineer techniques, to analyze the netlist disclosing sensitive information, such as cryptographic keys or algorithms.

M. Barbareschi (✉)
DIETI—Department of Electrical Engineering and Information Technologies, University of Naples Federico II, Via Claudio, 21 - 80125 Naples, Italy
e-mail: mario.barbareschi@unina.it

G. Di Natale · L. Torres
LIRMM UMR 5506—CNRS—University of Montpellier, 161 rue Ada, 34095 Montpellier Cedex 5, France
e-mail: giorgio.dinatale@lirmm.fr

L. Torres
e-mail: lionel.torres@lirmm.fr

N. Sklavos et al. (eds.), *Hardware Security and Trust*,
DOI 10.1007/978-3-319-44318-8_12

For these reasons, FPGA vendors have been starting to implement decryption algorithms on new and high-end FPGA devices, in order to program them by using enciphered bitstreams. Indeed, ciphered bitstreams guarantee confidentiality against IP theft and authenticity, such that it is not possible to use the bitstream on FPGAs that are not configured with the secret key. However, this technique is not a silver bullet for the FPGA security, since tampering and side-channel attack techniques are improving in efficacy and effectiveness, as recently demonstrated in [16].

With respect to the trustworthiness of integrated circuit (IC), the most important breakthroughs were given by the introduction of physically unclonable functions (PUFs) [5]. They exploit unavoidable and uncontrollable manufacturing imperfections, which are tolerated for the properly circuit operations, giving unique and unclonable hardware signatures. For instance, the propagation delay, through either nominally identical metal wires or through gates, depends on these variations. Hence, the PUF circuit has to mainly quantify a physical phenomenon affected by variability in order to be able to provide some responses. Since exploited quantities are from electrical phenomena, the responses are inherently affected by noise. The environmental and working conditions, such as the temperature and the supplied voltage, can dramatically alter PUFs responses, making them not suitable secure primitives due to lack of reliability.

PUFs work in a challenge/response paradigm, such that a PUF is a function which maps a set of inputs (challenges) to a set of outputs (responses) in a unique manner, defining a challenge/response pairs (CRPs) set. CRPs can be pragmatically used as key storage and key material provider and, if they are characterized by a huge cardinality, they can be even adopted in an authentication scheme [17]. PUFs are hard to attack and, furthermore, are tamper evident, such as physical attack attempts modify permanently their responses [5, 10].

Among all PUFs architectures that are discussed in the literature, we can list the SRAM PUF [4, 6], MRAM PUF [19] and the D flip-flop PUF [18] for the memory-based family, and the Arbiter PUF [8], the ring oscillator (RO) PUF [11–13, 17], the Butterfly PUF [7], and the Anderson PUF [2, 3] for the delay-based family. Ring oscillators-based PUFs (ROPUFs) are currently the most affordable secrecy source, since they can be easily implemented on every hardware technology, even on low-end and old FPGA device families, and received a great attention from the research community [13–15, 23, 24]. ROPUFs work by exploiting the variability on oscillations frequencies: considering a pair of ring oscillators (ROs), it is possible to extract one response bit by testing their frequencies with a binary comparator.

In this chapter, through a large amount of experiments conducted over Xilinx Spartan-6 XCS6LX16 45 nm devices, we collect some characterizations of RO frequencies, mainly aiming at analyzing how frequencies, generated by different ROs structures placed over a device and among different devices, are distributed. Along the way, we detail how to implement a RO on such devices and how to measure and extract frequencies from each implemented RO. In particular, for the frequency extraction we adopt Xilinx ChipScope. Furthermore, targeting a single device, we empirically study some noise sources, in particular the temperature variations, the logic which surrounds the ROs and the aging, in order to give better characterizations of read frequencies under different working conditions.

12.2 Xilinx FPGA Fabric

For any FPGA technology, the basic configurable element is the look-up table (LUT), which is a read-only memories able to implement any boolean function of k inputs depending on the configuration saved in it. Practically, the input of LUTs acts as an address signal and ROM values establish the output value of the function.

For Xilinx FPGAs, old and low-end devices are equipped with 4-input LUTs (e.g., Spartan-3/3E), while medium and high-end devices are characterized by 6-inputs LUTs. Any family arranges LUTs in more complex structures, called Slices. Most of them contains LUTs, flip-flops, carry propagation and generation logic and other *basic elements*. With the Spartan technology, Xilinx introduced the Slince M and L. Slice M is a Slice L with other additional features: indeed, Slice M has LUTs with can be configured to accomplish memory functionality, such as RAM, ROM, shift registers, and so on. Slices are grouped together to form the configurable logic block (CLB), which is strictly coupled with the switch matrix, used to communicate with other CLBs. CLBs are arranged in both spatial dimensions and this allows to address each CLB, and hence each slice, with two coordinates X and Y. Depending on the family, a CLB can include 2 or 4 slices, and typically each one contains one Slice M and one Slice L. Moreover, the Xilinx FPGA technology is characterized by a fast carry propagation path, that is a dedicated interconnection between slices that belong to the same columns, avoiding to route carry signals through switch matrices.

As for the Xilinx Spartan-6, adopted in this chapter, its fabric is characterized by 3 different slices [21]. Besides the previously described Slice L and slice M, The Slice X is the simplest structure as it is characterized by four 6-input LUTs and eight flip-flops. Each CLB of the Spartan-6 technology contains a pair of slices, either Slice L and Slice X or Slice M and Slice X. The two different slice pairs are alternated among CLBs columns, hence the odd CLBs columns contain a Slice L and Slice X pair, while the even columns are characterized by the pair Slice M and Slice X. The targeted FPGA device, namely the XC6SLX16, has a CLB array composed of 18 columns and 60 rows.

12.3 RO Frequencies Characterization

The ROPUF is an easily implementable hardware primitive and, with respect to other proposed PUFs architectures, it does not require special attention to symmetric placement, since its structure is a single closed loop [17]. For the FPGA technology, this implies a suitable implementation for every device and family. The design parameters which characterize the RO loop include: the number of stages, the routing and the placement of the loop. As for the first, it affects the oscillation frequency because the increase of involved stages in the loop causes a greater delay. In the same manner, the routing has to be considered since longer connections cause slower oscillation frequencies. At the end, the placement of the RO loop involves the choice of which basic elements implement the ring stages, i.e., the relative position among loop stages and the position of them in the chip.

Other parameters are able to alter the RO frequency, in particular the working conditions of the device, which are dynamic as they change over time, contrary to previous ones which are static and fixed at design time. Mainly, they can be considered as unwanted side effects and which cannot be controlled neither at design level nor during the lifetime. First of all, the supplied voltage is directly related to the signal propagation delay, hence the RO frequency is sensitive to the voltage variations. Those variations can be caused either by unstable supplied voltage or by variable workload of the logic that surrounds the RO, which absorbs a significant current and causes a local voltage drop. Hence, the switching activity, i.e., the logical values switching frequency of the signals of the surrounding logic represents also a disturb.

The die temperature, similarly to the voltage, is able to cause a degradation of the design performance, since at higher temperature values the signal propagation delay increases. Even in this case, two sources can be identified. Obviously the environmental temperature in which the circuit works is responsible for the signal delay over all the die, but a secondary contribution can be caused by local heating. Indeed, a high speed circuitry is able to warm the surrounding area due to the dissipation effect, hence it might affect the speed of other on-chip structures.

Another investigated side effect is the aging of the chip. Indeed, even with perfect and stable working conditions, during the chip lifetime, the frequency can be altered by the aging in a permanent manner. In fact, contrary to previous discussed effects, the aging of the chip is incremental and cannot be recovered once happened. Aging has different contributions, such as the hot carrier injection, the oxide breakdown, the electromigration phenomenon, the negative and positive bias temperature instability [9].

12.3.1 RO Structure and Measurement Architecture

The RO structure that we adopt in this chapter is reported in Fig. 12.1. A control gate interrupts the ring of inverters in order to enable or disable the oscillation; moreover its output is exploited to obtain the oscillating signal. If the inverting stages are odd, the control gate must be an AND gate, otherwise a NAND gate.

In order to measure the RO frequency value, we exploit two counters (see Fig. 12.2). One counter establishes the time window in which the frequency measurement has to be accomplished (the clock counter), hence it is timed by the system clock, which frequency is C, and it is configured to count up to a maximum fixed value T. The other, namely the ring oscillator counter, is fed with the RO output, so it

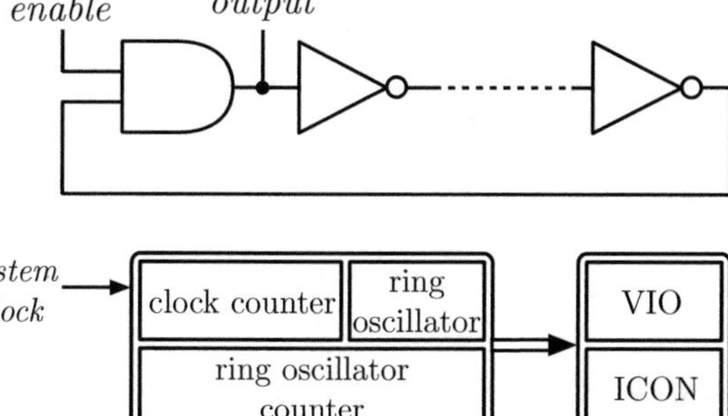

Fig. 12.1 Ring oscillator loop controlled by an AND gate

Fig. 12.2 High-level schematic view of the adopted design architecture

counts the edges (rising or falling) of the oscillations. When the first clock reaches the established maximum counting value, the RO is disabled and the RO frequency can be obtained computing $R \times C/T$, with R the number of counted oscillations edges. Due to the uncertainty on the system clock phases, this measurement introduces a measurement error $\varepsilon = \pm\frac{C}{2\times T}$.

The clock counter width is 18 bits and the RO counter width is fixed to 24 bits. This choice enables to keep the RO and the two counters bounded in a block of $6 \times 2 = 12$ CLBs, such that it can be easily placed over allowed FPGA positions. The resources allocated for the RO allows to instantiate a single RO with a number of stages up to 8.

As we are interesting in measuring frequency from ROs which are identically routed and placed on homologous structures, we defined each component as a relative placed macro (RPM). Contrary to hard macros, RPM allows to keep the original VHDL code and collect any information on the place and route as user constraint parameters. Each implemented entity is fixed thanks to a dedicated RPM in position within the cell and in used basic elements. This implies that each cell translation does not alter the design at the netlist level description. Furthermore, entities enriched with an RPM can be easily moved on the chip, arbitrary placing them anywhere.

As for the communication with a workstation to read the bits contained in the RO counter, the design architecture is instrumented with Xilinx ChipScope. ChipScope is a hardware and software suite that Xilinx includes in the design flow to allow the live debugging by means of the JTAG protocol. Indeed, by placing virtual I/O (VIO) and in logic analyzer (ILA), an user can instrument a design in order to provide inputs and retrieve outputs from a targeting HDL entity. Both the VIO and ILA are instantiated as IPCores within the design and they have to be properly connected with entity under test. Once the design flow is completed, the FPGA can be programmed with the obtained bitstream and analyzed by the ChipScope software, which offers a view pretty much like the HDL simulation environment. To retrieve the RO counter value, the design has to include only one VIO.

To automatize the frequency extraction phase, the Xilinx ChipScope have available a set of APIs which are accessible by means of the TCL scripts. So, by adopting such ChipScope libraries, the value can be retrieved by the workstation in which the board is plugged-in (Fig. 12.2).

Like the RO measurement cell, all involved ChipScope cores are fixed and bounded in an RPM.

12.4 Result and Validation

In this section, we show complete analysis for frequencies extracted from each implemented RO under different conditions. In particular, we analyze the effect of the logic which surrounds the RO, the temperature, and the aging. As for the static parameters, we analyze frequencies distribution over devices and with attention to the place and routing configurations.

12.4.1 Analysis of the Logic Surrounding the RO

As the surrounding logic is unavoidable to on-chip measure the frequency, it is necessary to evaluate how components surrounding ROs may influence their frequencies. At this aim, first of all we evaluated the impact of the proximity of both two counters and the ChipScope logic to the RO and, to avoid unwanted effects of temperature variations, we kept the external temperature fixed at 26.6 °C. This was accomplished by performing tests controlling the temperature of the FPGA by means of a thermal chamber. As for the RO, we targeted a single 5-stage RO and we implemented it by exploiting several syntheses, changing the on-chip position of the clock counter, RO counter and ChipScope, one by one keeping the others fixed. The design diversity allowed us to see how the surrounding logic involved in a frequency measurement affects read frequency values. Each experimental campaign involved about 1000 experiments and each one was repeated 25 times in order to mitigate the measuring error by averaging the values. Figure 12.3 reports frequencies distributions, obtained by allocating counters and the ChipScope logic in different positions, considered as percentage variations from the average value. As for the RO counter, its position does mostly not influence the frequency value, except for some positions around the same rows in which the RO is placed, causing an increase of 0.1 % on read values (Fig. 12.3a). In contrast, the read frequency is sensitive to the placement of the clock counter, with an alternating of decreases and increases of $-1\,\%/+\sim 3\,\%$ (Fig. 12.3b). In both cases, the measured frequencies turn out to be stable when the counters are placed close to the ROs. Figure 12.3c, d show that the impact of the ChipScope logic on read frequencies is practically insignificant, even when changing the shape in which its logic is bounded. In particular, placing ChipScope logic in different vertical positions does not have any significant effect, but moving it horizontally causes a slightly frequencies decrease (maximum ~0.05 %) proportionally

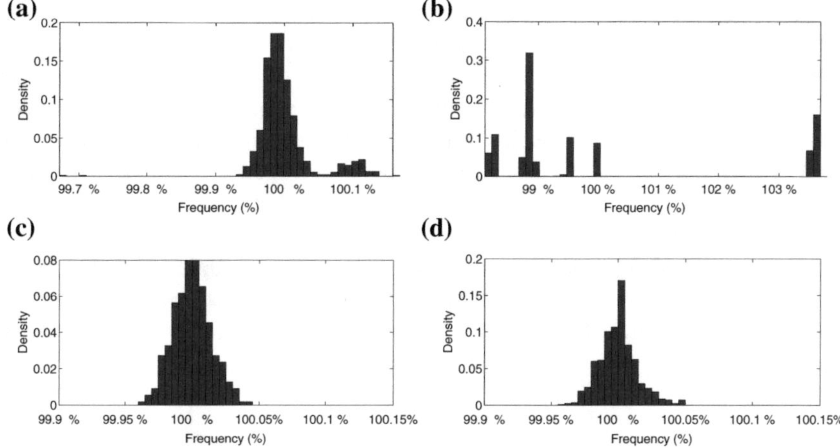

Fig. 12.3 Distribution of ROs frequencies values, considered as percentage variation from the average, with different places for counters and ChipScope debug logic

to the distance. Hence ChipScope can be considered as a nonintrusive surrounding logic. Indeed, during the oscillations sampling process, its logic does not work.

To better evaluate the effect of an intrusive logic that heavily works near the RO, we designed an architecture characterized by a pseudorandom behavior, inspired by the linear feedback shift register (LFSR). Compared to a classic LFSR, that is a single shift register whose serial input bit is a nonlinear function of previous states, we define a logic which perfectly fits the FPGA slices structure to guarantee higher workload. A very pervasive surrounding logic has to exploit all LUTs inputs with high switching activity signals and has to occupy as much as resources in slices in which it is allocated. In particular, each slice in the Spartan-6 fabric contains four 6-input LUTs and their outputs can be registered in flip-flops. Figure 12.4 shows

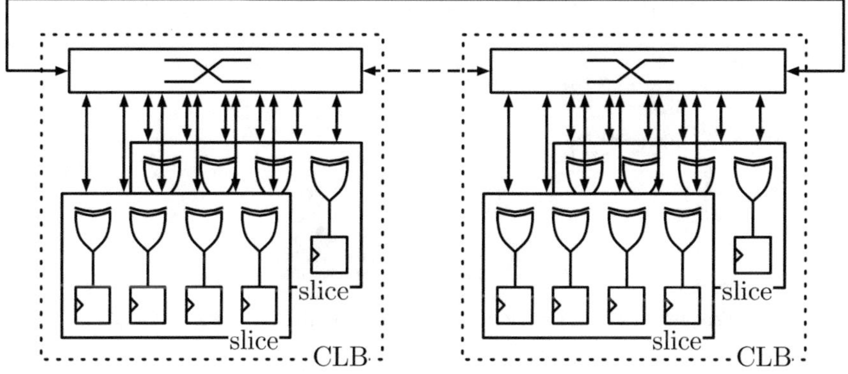

Fig. 12.4 A schematic overview of the implemented intrusive logic

a high level schematic of the logic that we designed. Each CLB has two slices which generate pseudorandom switching activity exploiting four parallel 6-input XOR functions and storing generated values in flip-flops. The input assignment for XOR function involves four signals locally picked, i.e., within the CLB, and 2 outside by neighbor CLBs. Iterating such a structure in a loop generates an auto-sustained signal switching, like the LFSR, but with more simultaneous activities per clock cycle.

The cells activity can be easily disabled, driving the signal `clock-enable` for all the involved flip-flops. The density of this intrusive logic, evaluated as the number of occupied LUTs on four times the number of occupied slices, reaches values between 75 and 85 %. Figure 12.5 reports three experiments with different

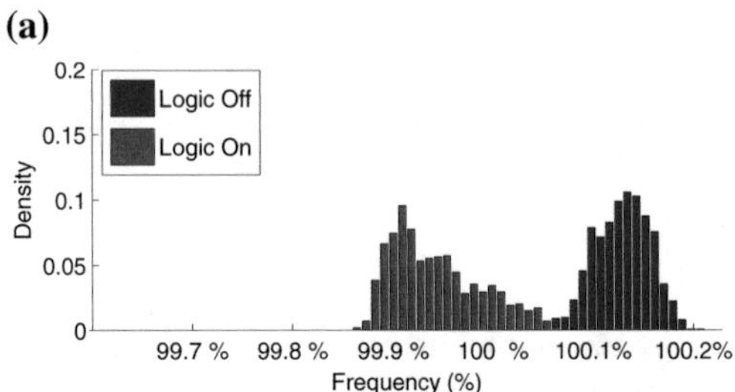

Frequencies distribution considering a shape of 3×8 CLB.

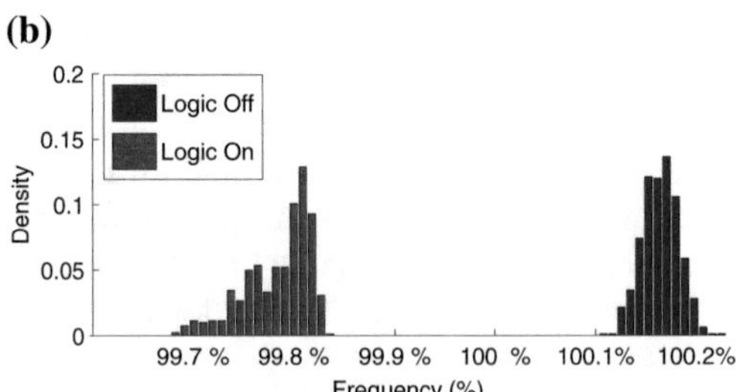

Frequencies distribution considering a shape of 8×6 CLB.

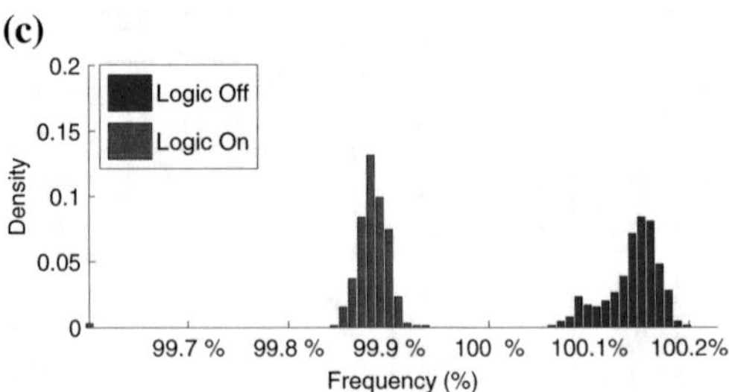

Frequencies distribution considering a shape of 6×18 CLB.

Fig. 12.5 Distribution of ROs frequencies values with an intrusive surrounding logic, considered as percentage variation from the value of ROs without the logic

intrusive logic configuration, respectively 3×8, 8×6 and 6×18, considering the frequency as percentage against average values of targeted RO measured without any insertion of intrusive logic. In all the reported cases, the logic causes a frequency decrease around 0.4 % when it is on and, surprisingly, it causes a frequency increase of about 0.15 % when it is off. Moreover, the logic is more intrusive in frequencies measurements when the prominent dimension is the height, hence the logic turns out to be more invasive if it is vertically stretched.

12.4.2 Analysis of the Stages Number and Routing

The frequency value of ROs is tightly coupled with the number of stages in the ring, such that longer loops cause lower frequencies. We designed five different loops with the number of inverting stages in the range between 4 and 8. In each design, the control gate of the loop was fixed in the bottom LUT of the Slice X and the loop has been arranged in the other available LUTs in slice X and the other slice, where its type (M or L) depends on the CLB column. Each RO measurement block is implemented in every allowable place on the FPGA, as in the previous experimental campaigns, obtaining about 1000 different design implementations. In Table 12.1 we report the means and standard deviations of the frequencies for the five different configurations.

The higher is the amount of stages, the lower are both average values and associated standard deviations. This implies that if the loop is longer, the values are closer to the average frequency values. Indeed, since each loop stage introduces a delay that is affected by uncertainty due to manufacturing variations, the uncertainty on the global delay turns out averaged.

A similar behavior can be also appreciated considering standard deviations of frequencies among different devices. In fact, we report in the same table two other standard deviations varying the number of RO stages: one is the average value, among 10 devices, of standard deviations calculated considering 10 different ROs, i.e., aver-

Table 12.1 Mean values and standard deviations of RO frequencies for different stages. The intra-die and inter-die are calculated among 10 ROs and 10 devices.

RO stages	Mean frequency (MHz)	Standard deviation (MHz)		
		Global	Intra-die	Inter-die
4	351.4552	7.1784	1.5295	2.5168
5	347.3760	7.0729	1.0850	2.0569
6	259.1042	5.1403	0.9113	1.4903
7	201.2785	3.6479	0.5729	0.9840
8	190.6770	3.7693	0.5320	0.8398

age intra-chip standard deviation; the other is the average value, among 10 ROs, of standard deviations calculated for 10 different devices, i.e., average inter-chip standard deviation. These two quantities represent how frequencies are scattered. In particular, the intra-die indicates the dispersion around the mean value for frequencies extracted from the same device, vice versa the inter-die measures the dispersion around the mean value considering frequencies read from different devices. Since intra-die standard deviations are greater than inter-die ones, it is clear that frequencies are closer among them when they are extracted from the same device rather than in the case in which they are measured from different devices. Moreover, standard deviations are inversely proportional to the number of stages. Furthermore, besides the RO structure, the placement and routing of the loop plays a crucial role in determining the oscillating frequencies. Aiming at deduce the relationship among different routing configurations and corresponding frequency changes, we have designed two identical 4-stage ROs but with different mapping within a Slice X. In particular, we have swapped two stages in the LUT assignment such that paths of the ring are different at least for two stages. The two configuration are allocated in every allowable location on the FPGA and their frequency distributions are reported in Fig. 12.6. The average values differ from one another by $\sim$100 MHz and their standard deviations by $\sim$2.35 MHz. Moreover, each configuration is characterized by a bimodal Gaussian as they are distributed around two well-distinguishable frequencies peaks. Correlating the frequencies with the spatial position, it is possible to note that ROs which are placed within a CLB in even columns, characterized by a slice X and slice L pair, have frequencies that in values are less than others placed in odd columns, characterized by a slice X and slice L pair. This happens even if the RO structure is entirely placed within a slice X. We can conclude that the two different distribution peaks are caused by different routing resources that characterize each CLB type.

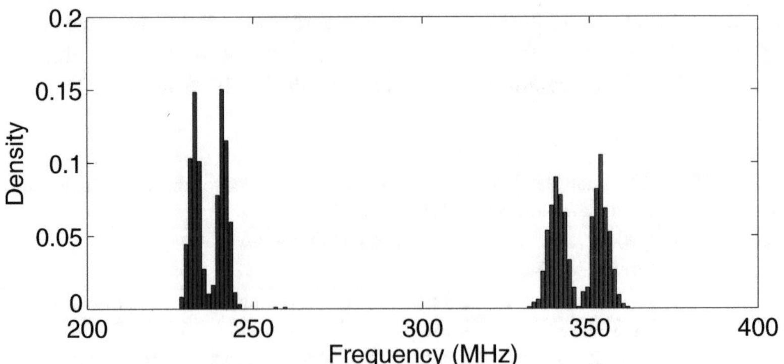

Fig. 12.6 Frequency distribution for two different mapping 4-stage RO configurations allocated over all the FPGA device

12.4.3 Temperature Analysis

The working temperature for an integrated circuit is directly related to the signal propagation delay. Indeed, high working temperatures cause a performance degradation in terms of speed. In order to analyze the effect of temperature changing on ROs frequencies, we measured frequencies of 5-stage ROs under 7 different external fixed temperatures, uniformly picked in the range between 0 and 80 °C, which is the commercial working temperature range [20]. To this aim, the FPGA was placed in a thermal chamber which keeps the temperature with a precision of 0.1 °C. Unfortunately, at 80 °C we were not able to correctly communicate with the FPGA in order to extract frequency value of the RO under test, hence we reported frequencies for the following values: 0, 13.3, 26.6, 40, 53.3 and 66.6 °C. When the thermal chamber reached the desired temperature, we waited 30 min in order to be sure that the die inside the package has uniformly reached the same external temperature before starting each test campaign. Figure 12.7 illustrates all measured frequencies varying the temperature with previous defined values between 0 and 66.6 °C. They are inversely proportional to temperature values and the relationship between them is quite close to be a linear function. The only observable exception is after 40 °C because the curve starts to be more descendant. In order to better analyze the relationship between the temperature value and the RO frequencies, we can consider the difference quotient for each temperature range, namely how the frequency decreases increasing the temperature of 13.33 °C. Figure 12.8 shows two distribution difference quotients calculated for all ROs. The blue histogram is related to the temperatures less than 40 °C, while the red to the temperature greater than 40 °C. They indicate that the average values of difference quotients are respectively −0.29 and −0.36 MHz/°C. Both are distributed with as a Gaussian curve with a standard deviation of ∼0.013 MHz/°C.

As for the even and odd CLB columns, there is a difference in terms of difference quotient, as illustrated in Table 12.2. The distance between the difference quotient of even and odd columns can be approximated as constant and equal to 0.0015 MHz/ °C.

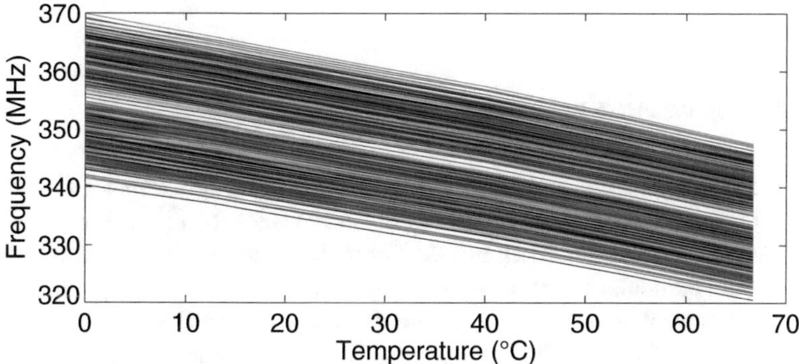

Fig. 12.7 Values for all ROs frequencies varying the working temperature

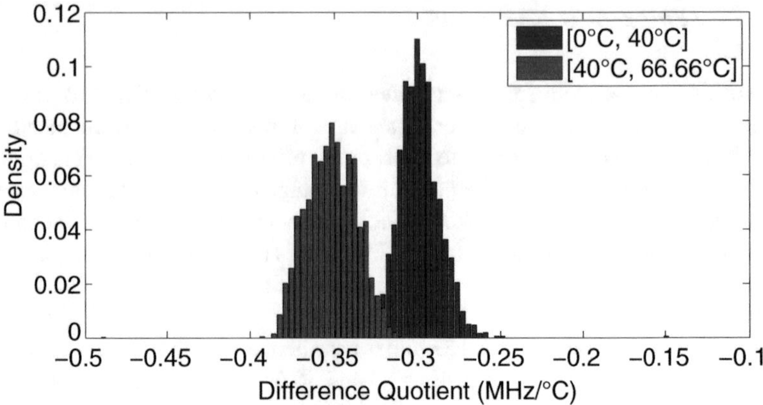

Fig. 12.8 Difference quotient distributions evaluated before and after 40 °C

Table 12.2 Frequencies and difference quotients for ROs placed in even and odd columns varying the working temperature

Temperature (°C)	Even column		Odd column	
	Frequency (MHz)	Difference quotient	Frequency (MHZ)	Difference quotient
0	348.5540		361.8935	
13.3	344.7655	−0.2842	357.9787	−0.2937
26.6	340.8063	−0.2970	353.8691	−0.3083
40	336.8726	−0.2951	349.7988	−0.3054
53.3	332.4376	−0.3327	345.1964	−0.3453
60	327.7191	−0.3540	340.2945	−0.3677

This implies that the frequencies of ROs placed in even columns are more sensitive than the ones placed in odd columns.

12.4.4 Aging Analysis

Aging is an unavoidable process that affects any IC, causing performance degradation and leakage current increase. In order to evaluate its impact on ROs, it is possible to perform aging acceleration of an FPGA. Through the application of stress working conditions, in particular high temperature and supplied voltage, for a period of time, the IC ages more than that period [1]. So we stressed a, FPGA core supplying an external voltage of 1.8 V (+50 % more than its nominal value [22]) and heating the chip up to 80 °C for 7 days. During this time, a particular design is configured onto the FPGA such that all the ROs are active at the same time. In order to clear the effects

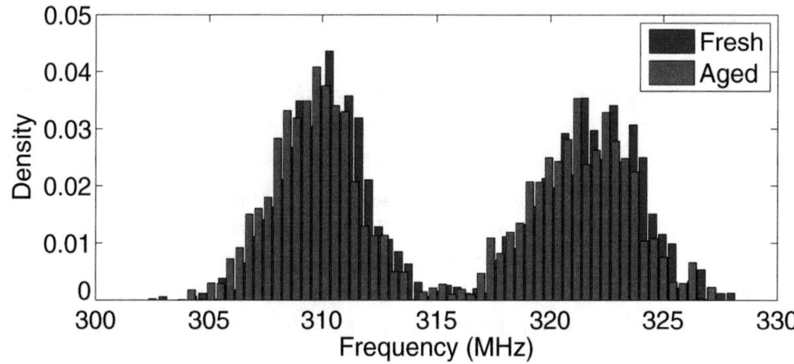

Fig. 12.9 Frequency distributions measured on the same device before and after the aging process

of the reversible aging process, we recovered the FPGA in 3 days with nominal supply voltage, but still keeping the temperature at 80 °C. As in the temperature experiments, we adopted 5-stage ROs to characterize the device before and after the aging process. Figure 12.9 illustrates such characterization through frequency histograms. As we can notice, the aged device frequency distribution has the same shape as the fresh version and it is shifted by ~0.6 MHz. In Fig. 12.10, we report the distribution of the difference quotients calculated on the frequencies of ROs freshed and aged. Even in this case, the values are distributed with a normal curve.

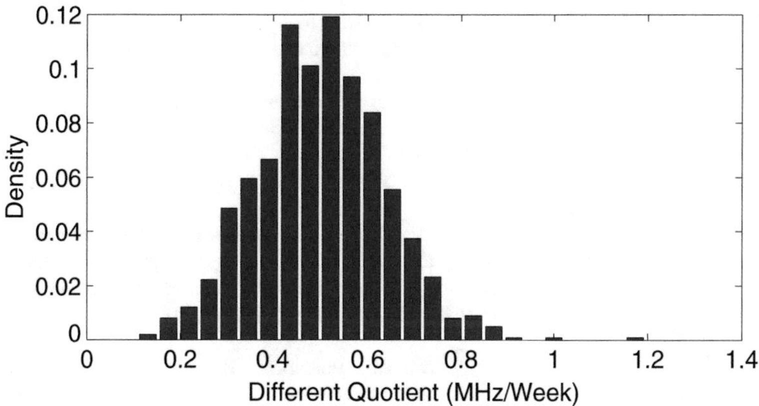

Fig. 12.10 Distribution of difference quotients of the aging campaign

12.5 Conclusion

In this chapter, we showed that ROs frequencies are tight coupled not only with design parameters, but also with other working conditions. In particular, we illustrated the role played by the on-chip logic which surrounds ROs and how the choice of the number of stages modifies ROs average frequencies and the dispersion of measured values around them. As for the working condition, we have posed our attention on the temperature, surrounding logic, and aging effects on the measured frequencies. Such analyses are extremely useful to design robust RO-based PUF, as we provide an extensive knowledge about the behavior of on-chip frequencies with static and dynamic parameters.

References

1. Amouri A, Bruguier F, Kiamehr S, Benoit P, Torres L, Tahoori M. Aging effects in fpgas: an experimental analysis. In: 2014 24th international conference on Field Programmable Logic and Applications (FPL); 2014. p. 1–4.
2. Anderson JH. A puf design for secure FPGA-based embedded systems. In: Proceedings of the 2010 Asia and South Pacific design automation conference. IEEE Press; 2010. p. 1–6.
3. Barbareschi M, Bagnasco P, Mazzeo A. Supply voltage variation impact on Anderson PUF quality. In: 2015 10th international conference on Design and Technology of Integrated Systems in Nanoscale Era (DTIS). IEEE; 2015. p. 1–6.
4. Barbareschi M, Battista E, Mazzeo A, Mazzocca N. Testing 90 nm microcontroller SRAM PUF quality. In: 2015 10th international conference on Design and Technology of Integrated Systems in Nanoscale Era (DTIS). IEEE; 2015. p. 1–6.
5. Gassend B, Clarke D, Van Dijk M, Devadas S. Silicon physical random functions. In: Proceedings of the 9th ACM conference on computer and communications security. ACM; 2002. p. 148–60.
6. Holcomb DE, Burleson WP, Fu K. Power-up SRAM state as an identifying fingerprint and source of true random numbers. IEEE Trans Comput. 2009;58(9):1198–210.
7. Kumar SS, Guajardo J, Maes R, Schrijen G-J, Tuyls P. The butterfly PUF protecting IP on every FPGA. In: IEEE international workshop on hardware-oriented security and trust, 2008. HOST 2008. IEEE; 2008. p. 67–70.
8. Lim D, Lee JW, Gassend B, Suh GE, Van Dijk M, Devadas S. Extracting secret keys from integrated circuits. IEEE Trans Very Large Scale Integr VLSI Syst. 2005;13(10):1200–5.
9. Lorenz D, Georgakos G, Schlichtmann U. Aging analysis of circuit timing considering NBTI and HCI. In: 15th IEEE international on-line testing symposium, 2009. IOLTS 2009. IEEE; 2009. p. 3–8.
10. Maes R, Verbauwhede I. Physically unclonable functions: a study on the state of the art and future research directions. In: Towards hardware-intrinsic security. Springer; 2010. p. 3–37.
11. Maiti A, Casarona J, McHale L, Schaumont P. A large scale characterization of RO-PUF. In: 2010 IEEE international symposium on Hardware-Oriented Security and Trust (HOST). IEEE; 2010. p. 94–9.
12. Maiti A, Schaumont P. Improving the quality of a physical unclonable function using configurable ring oscillators. In: International conference on field programmable logic and applications, 2009. FPL 2009. IEEE; 2009. p. 703–7.
13. Maiti A, Schaumont P. Improved ring oscillator PUF: an FPGA-friendly secure primitive. J Cryptol. 2011;24(2):375–97.

14. Merli D, Stumpf F, Eckert C. Improving the quality of ring oscillator PUFs on FPGAs. In: Proceedings of the 5th workshop on embedded systems security. ACM; 2010. p. 9.
15. Qu G, Yin C-E. Temperature-aware cooperative ring oscillator PUF. In: IEEE international workshop on hardware-oriented security and trust, 2009. HOST'09. IEEE; 2009. p. 36–42.
16. Skorobogatov S, Woods C. Breakthrough silicon scanning discovers backdoor in military chip. Springer; 2012.
17. Suh GE, Devadas S. Physical unclonable functions for device authentication and secret key generation. In: Proceedings of the 44th annual design automation conference. ACM; 2007. p. 9–14.
18. van der Leest V, Schrijen G-J, Handschuh H, Tuyls P. Hardware intrinsic security from D flip-flops. In: Proceedings of the fifth ACM workshop on scalable trusted computing. ACM; 2010. p. 53–62.
19. Vatajelu EI, Di Natale G, Barbareschi M, Torres L, Indaco M, Prinetto P. Spin-transfer torque magnetic random access memory (STT-MRAM). ACM J Emer Technol Comput Syst JETC. 2015.
20. Xilinx. Spartan-6 family overview. Available at http://www.xilinx.com/support/documentation/data_sheets/ds160.pdf.
21. Xilinx. Spartan-6 FPGA configurable logic block. Available at http://www.xilinx.com/support/documentation/user_guides/ug384.pdf.
22. Xilinx. Spartan-6 FPGA data sheet: DC and switching characteristics. Available at http://www.xilinx.com/support/documentation/data_sheets/ds162.pdf.
23. Xin X, Kaps J-P, Gaj K. A configurable ring-oscillator-based PUF for xilinx FPGAs. In: 2011 14th euromicro conference on Digital System Design (DSD). IEEE; 2011. p. 651–7.
24. Yin C-ED, Qu G. LISA: maximizing RO PUF's secret extraction. In: 2010 IEEE international symposium on Hardware-Oriented Security and Trust (HOST). IEEE; 2010. p. 100–5.

Index

N. Sklavos et al. (eds.), *Hardware Security and Trust*,
DOI 10.1007/978-3-319-44318-8

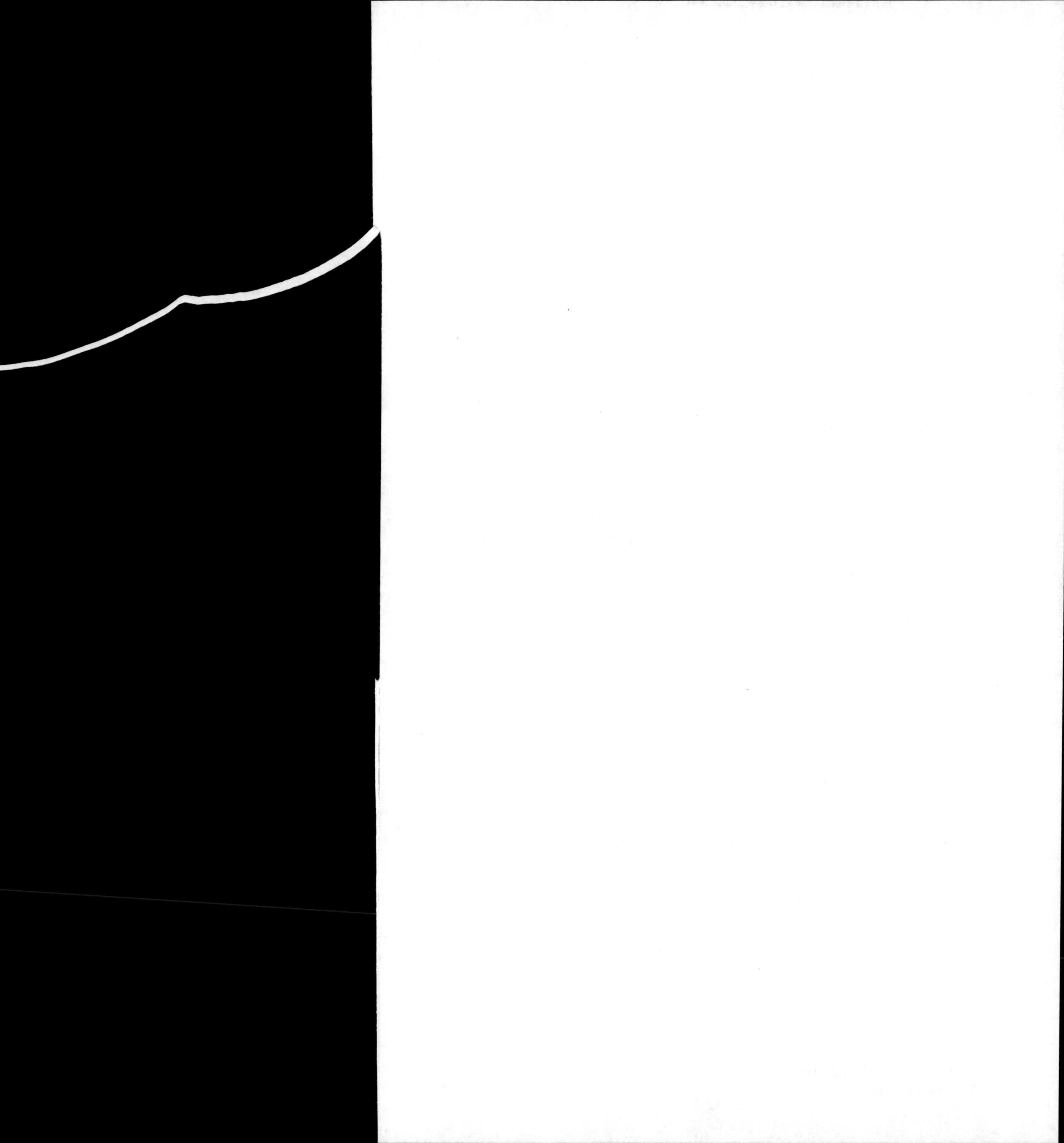

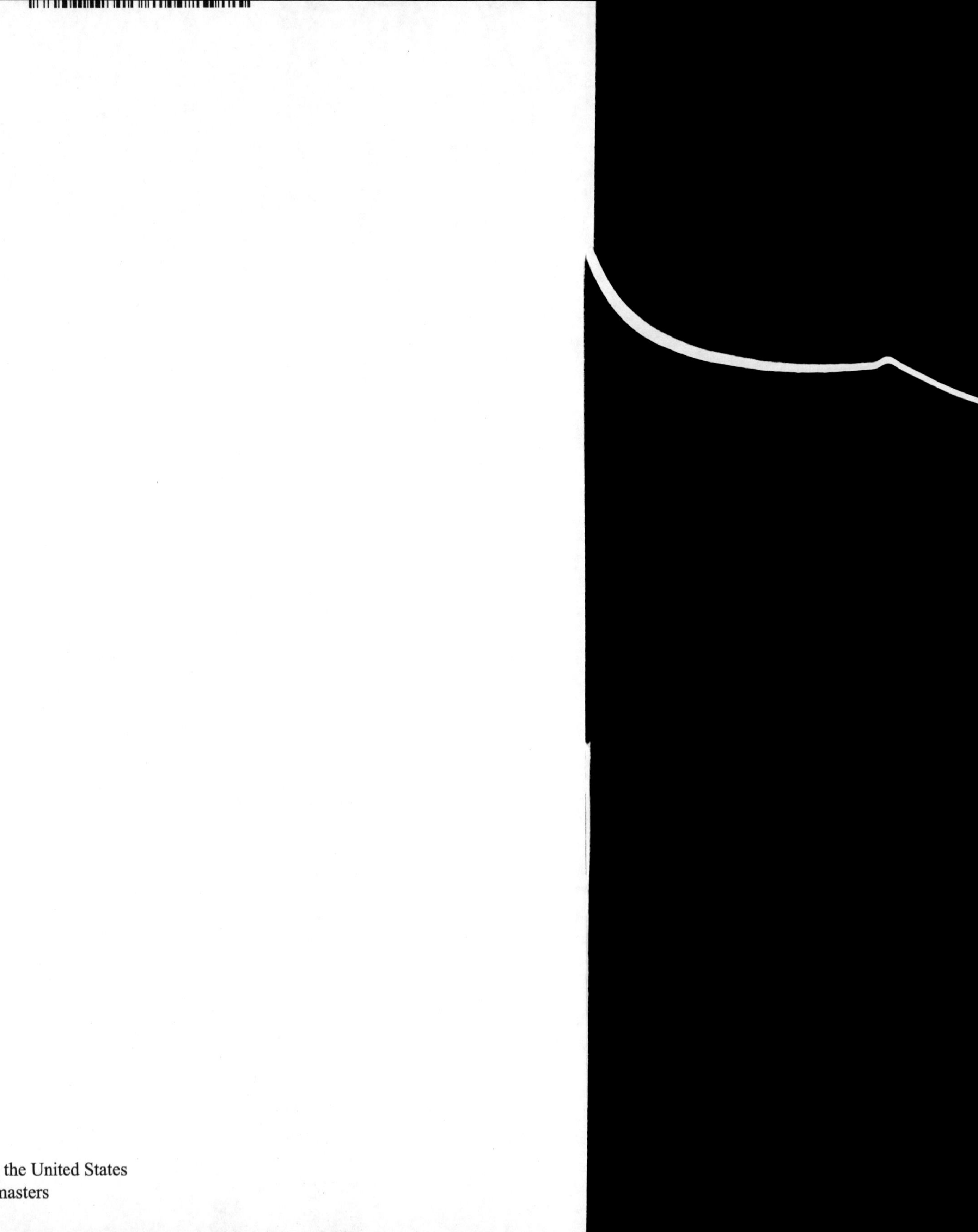

Printed in the United States
By Bookmasters